Soil Biological Fertility

Soil Biological Fertility

A Key to Sustainable Land Use in Agriculture

Edited by

Lynette K. Abbott

School of Earth and Geographical Sciences,
Faculty of Natural and Agricultural Sciences,
The University of Western Australia, Crawley, Australia

and

Daniel V. Murphy

School of Earth and Geographical Sciences,
Faculty of Natural and Agricultural Sciences,
The University of Western Australia, Crawley, Australia

 Springer

A C.I.P. Catalogue record for this book is available from the Library of Congress.

ISBN 978-1-4020-1756-8 (HB)
ISBN 978-1-4020-6618-4 (PB)
ISBN 978-1-4020-6619-1 (e-book)

Published by Springer,
P.O. Box 17, 3300 AA Dordrecht, The Netherlands.

www.springer.com

Printed on acid-free paper

Dedication

Professor Charles Alexander (Lex) Parker
1916-2001

Professor Lex Parker inspired many throughout his distinguished career. Lex was equally at home at the laboratory bench and in the paddock; his academic work was preceded by a period managing the family farm. He never lost touch with his farming background as he searched for scientific understanding of soil microbiological processes. He was renowned for exploring ideas that were not fashionable and was often ahead of his time. Lex commenced his academic career at The University of Western Australia in 1959 and was later appointed to a Personal Chair. He encouraged undergraduate students to pursue interests in soil and plant microbiology, and supervised a distinguished group of postgraduate students. Lex's research included studies of symbiotic and non-symbiotic nitrogen fixation, and he initiated the long running research on arbuscular mycorrhizas at The University of Western Australia. His subsequent research on root pathogens helped to highlight, then manage, the problems they caused in Western Australia, while inspiring others to address the problem nationally. During the later stages of his research career, Lex became passionate about soil fauna. He had loyal support from the farming community and persisted with his attempts to raise awareness of the importance of soil biology in agriculture. Many people greatly benefited from Lex's kind-heartedness, intuitiveness and enthusiasm.

TABLE OF CONTENTS

PREFACE

This book reviews the influences of management practices on soil biota and associated processes that contribute to soil biological fertility in the context of agricultural land use. Although it is generally acknowledged that physical, chemical and biological factors are all important to soil fertility (Figure 1), more attention is usually given to management of the soil chemical and physical environment than to the soil biological environment. Clearly, changes to the chemical and physical environment in soil influence biological processes and subsequently the contribution that they make to soil fertility overall. Certain soil biological processes are stimulated by soil amendments and this includes processes that have both positive and negative effects on plant growth. However, it is possible to override the contributions of some beneficial soil biological processes by solely focusing on chemical inputs to modify the chemical environment. Such an approach undervalues the potential for some key soil biological processes to contribute to plant productivity.

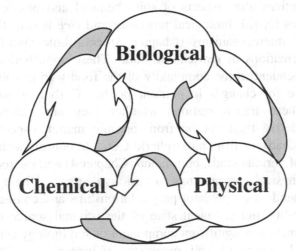

Figure 1 Soil biological, physical and chemical processes are interrelated.

A number of soil biological processes are linked, while others are not. Consideration of the mass of microorganisms in soil as a whole (microbial biomass) is not sufficient for complete interpretation of the effects of land management on soil biological fertility. When one group of organisms or biological processes is altered, others may be affected positively, negatively or not at all. Knowledge of these complex responses can support decision-making aimed at achieving sustainable use of agricultural land. However, the goals are not clearly defined as the biological characteristics of a particular soil type that is specifically managed to sustain the soil resource might be quite different for contrasting land uses such as food production, agroforestry or natural vegetation. Although the level of soil biological activity depends on the soil type, it also depends on the management practices used, particularly the management of organic matter.

Efficient use of nutrients requires a balance between those that are added to soil and those that are released during biological degradation of recent additions of plant/animal residues and older organic matter in soil. Aspects of soil biological activity that contribute to suppression of disease and efficiency of nutrient acquisition by plants are also essential for profitable and environmentally responsible agricultural production systems. Although management practices need to be appropriate for soil and climatic conditions if soils are to be sustained in a suitable biologically active state, many important questions remain unanswered before this can be achieved.

This book provides information about how agricultural land management practices alter aspects of soil chemical and physical fertility with consequences for soil biological processes and *vice versa*. Both fauna (Chapter 2) and microorganisms (Chapter 3) contribute significantly to chemical transformations of nutrients in soil. These contributions are not necessarily independent as the complexity of the food web in soil creates a dynamic interface for changes in nutrient pools. Carbon is an essential component in these transformations whether they are carried out by organisms that derive their carbon from organic matter, especially from plants (living or dead), or from atmospheric CO_2. In concert with chemical transformations of various kinds, both fauna (Chapter 4) and microorganisms influence their physical surroundings to various degrees. These processes are primarily dependent on predator-prey relationships and consequently on the form, availability and chemical state of the original source of carbon. The root environment is a significant component of soil biology and provides carbon to a wide range of soil organisms (Chapter 5). Rhizosphere organisms have major influences on plant nutrient availability and some, such as rhizobia, form specific associations with legumes (Chapter 6) which greatly influence the C:N ratio of plant material. Less specific, although almost ubiquitous associations between agricultural plants and arbuscular mycorrhizal fungi have the potential to increase the efficiency of use of phosphorus in agricultural ecosystems as well as improve soil structure (Chapter 7) but the extent to which is occurs in field soils is difficult to determine and disputed.

If agricultural management practices take account of biological processes in soil, there is a possibility of avoiding development of some severe plant diseases, especially those caused by root pathogens (Chapter 8). However, the complexity of soil biological processes is such that each process cannot be considered independently (Chapter 9). Therefore, an holistic approach to agricultural land management (Chapter 10) requires practices that are based on specific principles (Chapter 11) that ensure the whole range of important biological processes is not overridden. Only then can we develop sustainable farming systems for the future (Chapter 12). The goal is to take this knowledge and apply the principles to the improvement and modification of current farming practices on a localised and regional basis.

This book highlights the pivotal role of soil biology in agricultural ecosystems and demonstrates the responsiveness of soil biological fertility to changes in soil physical and chemical conditions imposed by agricultural management practices. A number of aspects of soil biology are not covered in depth here but they are certainly important to the holistic concept of soil biological fertility. They include the specific impacts of soil microorganisms on physical aspects of soil, the concept and importance of soil biodiversity, and the emerging new methodologies for investigating soil biodiversity and function.

Finally, we are extremely grateful to Merome Purchas for her valuable editorial assistance and general support throughout the preparation of this book.

Lynette K. Abbott
Daniel V. Murphy

June 2003

Chapter 1

What is Soil Biological Fertility?

Lynette K. Abbott and Daniel V. Murphy
School of Earth and Geographical Sciences, Faculty of Natural and Agricultural Sciences, The University of Western Australia, Crawley, 6009, WA, Australia.

1.   INTRODUCTION

There is increasing interest in soil management practices that enhance biological contributions to soil fertility due to greater awareness of the need for sustainable farming systems (e.g. Lynam and Herdt 1989, Dick 1992, Roper and Gupta 1995, Doran et al. 1996, Swift 1997, Condron et al. 2000, Mäder et al. 2002, von Lützow et al. 2002). This has occurred due to the requirement for better fertiliser use efficiency which is essential in: i) developing nations where cost and availability constrain production (Swift 1997) and   ii) many developed nations where public concern over environmental pollution from agricultural sources and associated government legislation restricts gaseous losses and nutrient leaching (e.g. N: Hatch et al. 2002, P: Leinweber et al. 2002). There has been a considerable decline in soil organic matter levels and associated loss of soil structure in many intensively cropped soils throughout the world. This has caused scientists

1

L.K. Abbott & D.V. Murphy (eds) Soil Biological Fertility - A Key to Sustainable Land Use in Agriculture. 1-15.
© 2007 *Springer.*

and landowners to consider more carefully how various components of the farming system can be managed to more efficiently benefit from biological processes that improve soil fertility.

Soil biological processes are extremely diverse and complex (Lavelle and Spain 2001). Physical and chemical soil characteristics, climate, plant communities and agricultural practices influence soil biology in a magnitude of ways, with both positive and negative influences on the overall fertility of soil. This level of complexity constrains our ability to assess or predict the biological state of soil through measures of abundance of organisms or their activity (Pankhurst *et al.* 1997). The current inability to predict the outcome of a change in agricultural management on soil biological processes, with a subsequent understanding of what this means in terms of production or the environment, is a major constraint to the successful design of farming systems that harness the biological potential of soil. Many studies have attempted to define the biological status of soil using simple indicator measurements (Doran and Parkin 1994, Gregorich *et al.* 1994, Franzluebbers *et al.* 1995, Pankhurst *et al.* 1995, Walker and Reuter 1996, Stenberg 1999). Whilst this is appealing to scientists, land holders and policy makers, it is extremely difficult to find correlations between potential indicators and crop production, long-term sustainability and environmental impact. Part of this difficulty has been with understanding how organisms and the functions that they perform interact with chemical and physical soil attributes in agricultural soils to regulate crop production and influence the longer-term status of the soil resource. Due to spatial and temporal heterogeneity and the enormous diversity displayed in soil biological characteristics, it is not easy to use them to define 'best practice' for land management.

The focus of this book is thus to provide an overview of a range of biological processes that contribute to soil fertility and to discuss the manner in which management practices influence soil biological fertility. With the complexity of these biological processes in mind, the impact of major management options and farming systems on soil biological processes can be addressed. The consequence of this is the basis for sustainable use of the whole soil resource, which demands equal consideration of biological, physical and chemical contributions to soil fertility. Inclusion of information about soil biological fertility in farm management decision-making should allow more precision in selecting inputs that complement the capacity of a soil to sustain production and minimise environmental damage such as might be caused by nutrient loss. If the type of production at a site is changed, different biological, physical and chemical states might be required to sustain the soil resource there, depending on the production system in place. A set of biological characteristics necessary for sustaining the soil resource at a particular site cannot be prescribed because different farming systems, or even different stages in the same farming system, might require differences in soil biological fertility. Therefore, a suite of soil biological characteristics needs to be defined for each land use category according to the soil type and climatic conditions.

2.   WHAT IS SOIL BIOLOGICAL FERTILITY?

There has been a great deal of discussion about the use of terms to describe the state of soil - e.g. soil quality, soil fertility, and soil health - as a means of improving recognition of the importance of the soil resource. In an agricultural context, the historical term - soil fertility - has the ability to convey all of the qualities required for plant and animal production. Soil has usually been investigated primarily from the perspective of pedological, physical, chemical and hydrological characteristics. This is the case even though soil organisms mediate a number of important pedological, physical and chemical processes (Lavelle and Spain 2001). The concept of soil fertility has generally been most concerned with soil chemical fertility and its ability to meet the nutritional needs of plants. For chemically-based farming systems, fertiliser requirements can be determined according to plant, soil and climatic conditions and extensive research has been carried out to identify these requirements in many agricultural situations. The physical constrains to soil fertility are also widely acknowledged and considerable effort has been expended in identifying land use practices that prevent or minimise development of structural constraints to plant growth or to soil loss through wind or water erosion. In contrast, much less is known about i) how to maximise benefits from soil biological processes (with the exception of symbiotic N fixation and biological control of plant disease), and ii) whether it is economically or environmentally sustainable to capture benefits from other soil biological processes.

The term 'soil fertility' used without the qualifiers 'biological', 'physical' or 'chemical' gives insufficient information about the state of soil. These three prefixes allow interpretation to be focused on components, or combinations of components, of soil fertility that are influenced by management decisions. Soil biological fertility has been used in this book in preference to terms such as 'soil biological quality' and 'soil biological health' within the framework set out in Table 1. Unfortunately, there are no simple, widely applicable and quantitative measures of any of the aspects of soil biological fertility because they are constrained by parent rock, soil origin, landscape and climatic factors as well as by land use. In spite of this, we recommended that the term *soil biological fertility* become widely used with reference to agricultural production systems. Without a focus on this component of soil fertility, the contributions of beneficial soil biological processes will continue to be consumed within the context of physical and chemical fertility and not recognised as an equally important aspect of the soil resource.

Table 1 Suggested working 'definitions' of soil fertility and its components: soil biological fertility, soil chemical fertility and soil physical fertility. The terms only have general conceptual significance because they cannot be quantified exactly or defined in specific units. For a particular site, the 'degree' of soil fertility (and components of soil fertility) depends on the inherent characteristics of the soil according to its origin and on the land management practices implemented.

COMPONENT OF SOIL FERTILITY	'DEFINITION'
SOIL FERTILITY	The capacity of soil to provide physical, chemical and biological requirements for growth of plants for productivity, reproduction and quality (considered in terms of human and animal wellbeing for plants used as either food or fodder) relevant to plant type, soil type, land use and climatic conditions.
SOIL BIOLOGICAL FERTILITY	The capacity of organisms living in soil (microorganisms, fauna and roots) to contribute to the nutritional requirements of plants and foraging animals for productivity, reproduction and quality (considered in terms of human and animal wellbeing) while maintaining biological processes that contribute positively to the physical and chemical state of soil.
SOIL CHEMICAL FERTILITY	The capacity of soil to provide a suitable chemical and nutritional environment for plants and foraging animals for productivity, reproduction and quality (considered in terms of human and animal wellbeing) in a way that supports beneficial soil physical and biological processes, including those involved in nutrient cycling.
SOIL PHYSICAL FERTILITY	The capacity of soil to provide physical conditions that support plant productivity, reproduction and quality (considered in terms of human and animal wellbeing) without leading to loss of soil structure or erosion and supporting soil biological and chemical processes.

3.　IMPORTANCE OF SOIL BIOLOGICAL FERTILITY TO AGRICULTURAL PRODUCTION

If the fertility of soil is considered in terms of short-term agricultural production alone, there may be little need for attention to soil biological processes in many developed nations where soils are inherently well supplied with major nutrients (for global soil nutrient maps see Figure 3, Huston 1993). This is because many of the benefits provided by soil organisms can be overridden by the indigenous nutrient supply or by the addition of synthetic fertilisers where inorganic nutrients and chemicals are readily available and relatively inexpensive. Biological processes that are exceptions to this are plant disease and symbiotic nitrogen fixation, which can both have significant effects on production in predominantly chemical-based agricultural systems. Generally, the emphasis of 'modern' agriculture, with widespread introduction of synthetic fertilisers, has largely ignored the potentially beneficial contributions of some soil organisms. This approach has lead to serious contamination of some environments by pesticides and nutrients including nitrogen, phosphorus or even trace elements such as copper. Furthermore, modern plant varieties have often been selected under conditions that are not favourable for certain biological processes (such as the function of arbuscular mycorrhizas (Smith *et al.* 1993)). This might create agricultural environments that cause some potentially positive aspects of soil biological fertility to be detrimental (Ryan and Graham 2002).

A pedological context to soil biology has been presented in great detail by Lavelle and Spain (2001). It provides a necessary perspective for evaluation of the importance of soil biological processes and for identifying underlying principles that can be applied across soil types and environments. It cannot be assumed that soil biological processes are effective unless demonstrated to be so for specific environmental and soil conditions. Furthermore, land management practices (such as fertiliser use) can alter soil conditions substantially to facilitate growth of agricultural plants that are not naturally suited to the original soil conditions. Other changes in both physical and chemical conditions might be able to be mediated by soil organisms if they are provided with an energy source (e.g. from manure, mulching or stubble retention).

Although larger organisms such as earthworms and termites can substantially influence the structural characteristics of soil, the greatest impact of smaller organisms is likely to be on soil chemical characteristics. Some chemical and physical processes in soil have significant influences on one another independently of biological processes. Thus, the interdependent nature and complexity of soil processes means that a simplistic view to its assessment is not appropriate.

It is not routine practice to prescribe soil biological conditions suited to the needs of individual farming systems at specified locations, although this is attempted with organic farming practices (Stockdale *et al.* 2001). Different sets of soil biological attributes may be more or less appropriate as farming practices are changed. If a soil has a high content of nitrogen or phosphorus (in terms of the adequacy of these nutrients for plant growth) it would probably be considered to be 'fertile'. However, this level of chemical fertility could have been derived primarily from synthetic fertiliser inputs. Alternatively, substantial contributions may have come from processes involving interactions between organisms, decomposition of organic matter and cycling of nutrients, or from a combination of organic and inorganic inputs and biological processes related to organic matter degradation. If soil organisms were major contributors to the high level of chemical fertility (through their interactions with organic matter inputs), corresponding positive contributions to the physical state of soil would most likely result. Furthermore, the biological processes and chemical inputs that contribute to soil chemical fertility can be linked. For example, higher fertiliser input leads to increased plant biomass that can enhance nutrient cycling through soil organisms if the organic matter has the required elemental content and is managed appropriately. The capacity of the soil to retain these nutrients is of great importance to their efficient use for agricultural production and to ensure that there is no loss into the environment through leaching and other means of dispersion.

From the perspective of enhancing and/or preserving the soil resource, the effect on the soil of increases in chemical fertility arising from either a biological or a chemical source may be quite different. However, evaluation of the biological component of soil fertility has generally been considered unnecessary when 'available' nutrient levels are 'adequate'. The evaluation of components of soil biological fertility presented in this book combine to demonstrate the breadth of contributions that soil biological processes make to the state of soil (Table 2).

4.   MEASUREMENT OF SOIL BIOLOGICAL FERTILITY

There are well-established criteria for defining 'ideal' conditions of both soil chemical and physical fertility across diverse soils of different origins and in different climatic zones (e.g. Karlen and Stott 1994, Cass *et al.* 1996, Merry 1996). However, there is a lack of fundamental understanding of how soil biological, chemical and physical soil attributes interact and how they

Table 2 Examples of how management options influence soil biological processes and change soil fertility.

Management option	Soil biological processes influenced	Change to soil chemical fertility	Change to soil physical fertility
Organic matter Crop residue, farmyard and green manure incorporation, including composted household or green waste material.	Microbial populations enhanced. Initially may cause immobilisation of soil nutrients due to high C:N of organic residues. Soil faunal groups may increase or decrease depending on residue quantity and quality.	Increased availability of nutrients especially N, S and P. Cation exchange capacity increased (significant in low clay soils). Some potential for accumulation of heavy metals.	Improve soil structure through aggregation and water holding capacity.
Crop rotation Use of pastures including legumes.	Increased abundance of soil fauna, in particular earthworms.	C:N ratio decreased.	Preferential flow paths improving aeration and water infiltration.
Crop rotation Mixed plant species cropping.	Greater range of root exudates supports diverse microbial population.	Changes in C:N ratio.	Increased localised heterogeneity of soil.
Tillage Minimum/no tillage.	Encourages re-colonisation by soil faunal species and functional groups. Increase in overall faunal abundance and encourages greater diversity. Alters the bacterial:fungal ratio favouring fungal feeding faunal species. Concentrates soil microbial biomass into surface soil layer.	Decreased soil organic matter decomposition rates initially cause nutrient immobilisation. However, build up of soil organic matter results in greater net nutrient release.	Improved soil structure through aggregation and water holding capacity. Improved macro-pore formation.

Table 2 Continued on Page 8

Table 2 Continued

Management options	Soil biological processes influenced	Change to soil chemical fertility	Change to soil physical fertility
Controlled traffic	Minimises soil compaction resulting in higher soil faunal abundance and an increase in microbial biomass.	Increased availability of nutrients.	Improved total porosity leading to greater aeration and better water infiltration.
Livestock grazing	Provides a C-source, enhancing microbial and faunal activity, unless increased soil compaction decreases microbial activity.	Reduced mineralisation due to increased compaction and less organic matter.	Increased soil compaction.
Fertiliser High application rates	High P reduces mycorrhizal colonisation of roots. High N inhibits N_2 fixation. Nitrifier populations increased with NH_4^+ fertiliser. Some soil faunal groups increased.	Decreased availability of nutrients derived from soil biological processes. Soil acidification. Increased nitrate leaching to groundwater.	
Inoculants *Rhizobium* inoculation	N_2 fixation from the atmosphere. Introduction of more effective strains dependent on indigenous populations.	Increased N availability in soil on subsequent decomposition of legumes. Soil acidification.	
Inoculants *Penicillum radicardium*	Mineralisation of non-soluble P rich fertiliser products.	Increase P available to plant.	
Pesticides	Some beneficial soil fauna also killed. Potential to loose beneficial species completely, thus altering food webs.	Nutrient supply altered depending on shift in food web.	

are changed by agronomic management practices. Biological processes often have an indirect effect on plant growth (e.g. via nutrient availability or soil structure) making it difficult to illustrate a benefit to crop production. More than one combination of soil biological properties could be considered ideal, so it is difficult to define an optimal biological state of soil or the precise importance of biodiversity of organisms in agricultural soils.

Development of more quantitative research techniques to estimate biodiversity of organisms in soil and the dynamics of nutrient pools mediated by organisms have enabled specific management practices and more complex farming systems to be studied in ways not previously possible. For example, the development of techniques for assessing nutrients in the soil microbial biomass was a major advance for the rapid and routine study of soil biological processes associated with organic matter (Jenkinson and Powlson 1976-a, 1976-b, Brookes *et al.* 1982, Brookes *et al.* 1985). The capacity to quantify the mass of the bacterial and fungal population (compared to direct microscopy and plating techniques) played an important role in advancing knowledge of the dynamics of organic matter breakdown (Powlson and Brookes 1987) and associated nutrient cycling (Jenkinson and Parry 1989). In recent years, there has been a major advance in the assessment of more specific biochemical, functional and molecular characteristics of soil biology (Torsvik *et al.* 1990-a, 1990-b, Turco *et al.* 1994, Zak *et al.* 1994, Degens and Harris 1997, Tiedje *et al.* 2001, Murphy *et al.* 2003). These advances have allowed focus to shift from determination of types of organisms present to an assessment of the contribution of biological processes to key beneficial soil functions. The focus on identifying functional diversity of soil communities (Lupwayi *et al.* 1998, Kennedy 1999, Altieri 1999) provides the opportunity to determine causal effects on plant production and longer-term predictions of the future soil status.

In parallel with improved technology for assessing components of communities of soil organisms, emphasis has been placed on the importance of sampling strategies including time of sampling, depth of sampling, spatial distribution, storage of samples and use of volumetric units of measurement (Doran *et al.* 1996, Glendining and Poulton 1996, Sparling 1997, Degens and Vojvodi-Vukovi 1999, Shi *et al.* 2002, Smith *et al.* 2002). Although many technical advances have been made and new methods have become available for the assessment of specific components of soil biological fertility, it is essential to ensure that they are suitable to the soil conditions where they are used (e.g. Murphy *et al.* 2003). Inappropriate use of this technology, such as use without regard for local soil conditions, will lead to confusion and misinformation about soil biological fertility in relation to farming systems. The interpretation of data related to soil biological fertility remains an impediment to the development and implementation of models of nutrient cycling.

5.   APPLYING KNOWLEDGE OF SOIL BIOLOGICAL FERTILITY TO FARMING PRACTICES

The complex nature of biological processes in soil is well recognised and it is not possible to characterise in detail the whole of the soil biology at every site. This means that day-to-day recommendations for improving the sustainability of farming systems are very seldom based on well-defined (if any) measures of soil biological fertility. Soil biological fertility is dynamic and quantitative measures vary greatly with time, even within short periods. The heterogeneity of soil biological processes in soil (Strong *et al.* 1998) presents further difficulty for quantification of soil biological fertility. Yet another problem is the conflicting views of what constitutes an ideal value. Crop production, long-term soil sustainability and environmental concerns often require opposing classifications of what is an acceptable indicator value (Sojka and Upchurch 1999). Therefore, the concept of defining acceptable and critical values for soil biological indicators has not been successful (Sojka *et al.* 2003). Thus 'one-off' measurements are not particularly useful for characterising the biological status of a soil. This contrasts with measures of other soil characteristics, such as pH, which change relatively slowly over time and allow 'one-off' measures to be applicable beyond the time of sampling.

Measurements of specific aspects of soil biology can be successfully applied to the comparison of management practices (e.g. tillage versus no tillage) or contrasting farming systems (e.g. organic versus conventional). Although measurements of soil biological characteristics are often difficult to interpret, their advantage over chemical and physical characteristics is that they are often more responsive to changes in management practice (Figure 1). For example, microbial biomass and biologically active fractions of soil organic matter turnover within months to a few years (Jenkinson and Ladd 1981) whilst the majority of soil organic matter takes decades or longer to turnover (Stout *et al.* 1981). Although it is generally not possible to define an optimal value for microbial biomass in a soil, if the microbial biomass or ratio of microbial biomass-carbon to total-carbon increased, this would be perceived as an improvement to the soil (Sparling 1997) even though it may not be expressed in terms of nutrient availability, plant production or yield (Fauci and Dick 1994, Sorn-srivichai *et al.* 1988). For this reason, such measurements are often well suited to monitoring programs where the emphasis is on assessment of the change in direction of a soil characteristic over time. Soil biological characteristics that change rapidly could be useful indicators of the impacts of agricultural practices. The current fundamental understanding of the importance of these characteristics to soil conditions can be used to make valued judgements as to the importance of the degree and direction of change of the indicators in response to agricultural practice.

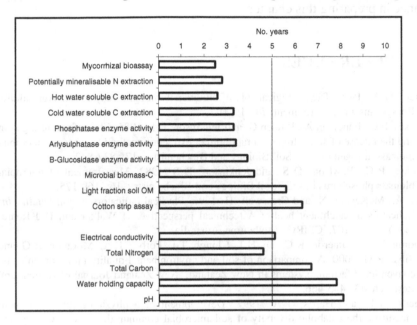

Figure 1 Number of years required to detect significant differences in soil measurements between management practices assessed on a grey clay soil (Sodosol) in south-western Australia (Milton et al. 2002). Land managers have a low uptake rate for soil monitoring (Lobry de Bruyn and Abbey 2003) and factors that take more than five years before a change can be detected are likely to have little impact on their decision-making.

Although soil biological characteristics can be monitored, this does not overcome the difficulty of knowing if they are either within an acceptable range or over an acceptable threshold value if one does indeed exist. Furthermore, the rate of change of the measured soil parameter may provide more insight into the impact of management on soil biological fertility than the magnitude of the parameter *per se*. Fundamental understanding of how specific soil biological characteristics respond to management practices is required if the characteristics are to be used as indicators. More importantly, information is required about the relationship between the biological characteristics and plant production, development of a sustainable soil matrix and/or prevention of environmental problems.

6.　ACKNOWLEDGEMENTS

We have received valuable support from the Grains Research and Development Corporation, the Rural Research and Development Corporation and The University of Western Australia for research on soil biological fertility. Finally, we thank Nui Milton and Merome Purchas for their assistance in preparing this chapter.

7.　REFERENCES

Altieri M A, 1999 The ecological role of biodiversity in agroecosystems. Agriculture, Ecosystems and Environment 74: 19-31.

Brookes P C, Landman A, Pruden G and Jenkinson D S 1985 Chloroform fumigation and the release of soil nitrogen: a rapid direct extraction method to measure microbial biomass nitrogen in soil. Soil Biology and Biochemistry 17: 837-842.

Brookes P C, Powlson D S and Jenkinson D S 1982 Measurement of microbial biomass phosphorus in soil. Soil Biology and Biochemistry 16: 169-175.

Cass A, McKenzie N and Cresswell H 1996 Physical indicators of soil health. *In:* Indicators of catchment health: A technical perspective. J Walker and D J Reuter (eds.) pp. 89-107. CSIRO. Melbourne, Australia.

Condron L M, Cameron K C, Di H J, Clough T J, Forbes E A, McLaren R G and Silva R G 2000 A comparison of soil and environmental quality under organic and conventional farming systems in New Zealand. New Zealand Journal of Agricultural Research 43: 443-466.

Degens B P and Harris J A 1997 Development of a physiological approach to measuring the catabolic diversity of soil microbial communities. Soil Biology and Biochemistry 29: 1309-1320.

Degens B P and Vojvodi -Vukovi M 1999 A sampling strategy to assess the effects of land use on microbial functional diversity in soils. Australian Journal of Soil Research 37: 593-601.

Dick R P 1992 A review: Long-term effects of agricultural systems on soil biochemical and microbial parameters. Agriculture, Ecosystems and Environment 40: 25-36.

Doran J W and Parkin T B 1994 Defining and assessing soil quality. *In:* Defining soil quality for a sustainable environment. J W Doran, D C Coleman, D F Bezdicek and B A Stewart (eds.) pp. 3-21. Soil Science Society of America Special Publication No 35. Madison , WI.

Doran J W, Sarrantonio M and Liebig M A 1996 Soil health and sustainability. Advances in Agronomy 56: 1-54.

Fauci M F and Dick R P 1994 Microbial biomass as an indicator of soil quality: Effects of long-term management and recent soil amendments. *In:* Defining soil quality for a sustainable environment. J W Doran, D C Coleman, D F Bezdicek and B A Stewart (eds.) pp. 229-234. Soil Science Society of America Special Publication No 35. Madison , WI.

Franzluebbers A J, Zuberer D A and Hons F M 1995 Comparison of microbiological methods for evaluating quality and fertility of soil. Biology and Fertility of Soils 19: 135-140.

Glendining M J and Poulton P R 1996 Interpretation difficulties with long-term experiments. *In:* Evaluation of soil organic models. NATO ASI Series I, Vol 38. D S Powlson, P Smith and J L Smith (eds.) pp. 99-109. Springer-Verlag. Berlin.

Gregorich E G, Carter M R, Angers D A, Monreal C M and Ellert B H 1994 Towards a minimum data set to assess soil organic matter quality in agricultural soils. Canadian Journal of Soil Science 74: 367-385.

Hatch D, Goulding K and Murphy D 2002 Nitrogen. *In:* Agriculture, Hydrology and Water Quality. P M Haygarth and S C Jarvis (eds.) pp. 7-27. CAB International. Wallingford, Oxon. UK .

Huston M 1993 Biological diversity, soils and economics. Science 262: 1676-1680.

Jenkinson D S and Ladd J N 1981 Microbial biomass in soil: Measurement and turnover. *In:* Soil Biochemistry Vol 5. E A Paul and J N Ladd (eds.) pp. 415-471. Marcel Dekker. New York.

Jenkinson D S and Parry L C 1989 The nitrogen cycle in the Broadbalk wheat experiment: A model for the turnover of nitrogen through the soil microbial biomass. Soil Biology and Biochemistry 21: 535-541.

Jenkinson D S and Powlson D S 1976 (a) The effects of biocidal treatments on metabolism in soil - V. A method for measuring soil biomass. Soil Biology and Biochemistry 8: 209-213.

Jenkinson D S and Powlson D S 1976 (b) The effects of biocidal treatments on metabolism in soil – I. Fumigation with chloroform. Soil Biology and Biochemistry 8: 167-177.

Karlen D L and Stott D E 1994 A framework for evaluating physical and chemical indicators of soil quality. *In:* Defining soil quality for a sustainable environment. J W Doran, D C Coleman, D F Bezdicek and B A Stewart (eds.) pp. 53-72. Soil Science Society of America Special Publication No 35. Madison, WI.

Kennedy A C 1999 Bacterial diversity in agroecosystems. Agriculture, Ecosystems and Environment 74: 65-76.

Leinweber P, Turner B L and Meissner R 2002 Phosphorus. *In:* Agriculture, Hydrology and Water Quality. P M Haygarth and S C Jarvis (eds.) pp. 29-55. CAB International. Wallingford, Oxon. UK.

Lavelle P and Spain A V 2001 Soil Ecology. Kluwer Academic Publishers. Dordrecht/Boston/London. pp 654

Lobry de Bruyn L A and Abbey J A (2003) Characterisation of farmers' soil sense and the implications for on-farm monitoring of soil health. Australian Journal of Experimental Agriculture 43: 285-305.

Lupwayi N Z, Rice W A and Clayton G W 1998 Soil microbial diversity and community structure under wheat as influenced by tillage and crop rotation. Soil Biology and Biochemistry 30: 1733-1741.

Lynam J K and Herdt R W 1989 Sense and sustainability: Sustainability as an objective in international agricultural research. Agricultural Economics 3: 381-398.

Mäder P, Fliessbach A, Dubois D, Gunst L, Fried P and Niggli U 2002 Soil fertility and biodiversity in organic farming. Science 296: 1694-1697.

Merry R H 1996 Chemical indicators of soil health. *In:* Indicators of catchment health: A technical perspective. J Walker and D J Reuter (eds.) pp. 109-119. CSIRO. Melbourne, Australia.

Murphy D V, Recous S, Stockdale E A, Fillery I R P, Jensen L S, Hatch D J and Goulding K W T 2003 Gross nitrogen fluxes in soil: Theory, measurement and application of ^{15}N pool dilution techniques. Advances in Agronomy 79: 69-118.

Milton N, Murphy D, Braimbridge M, Osler G, Jasper D and Abbott L 2002 Using power analysis to identify soil quality indicators. 17th World Congress of Soil Science Symposium No. 32, Paper No. 557 pp. 1-8

Pankhurst C E, Doube B M and Gupta V V S R (eds.) 1997 Biological indicators of soil health. CAB International. Wallingford, Oxon. UK.

Pankhurst C E, Hawke B G, McDonald H J, Kirkby C A, Buckerfield J C, Michelsen P, O'Brien K A, Gupta V V S R and Doube B M 1995 Evaluation of soil biological properties as potential bioindicators of soil health. Australian Journal of Experimental Agriculture 35: 1015-1028.

Powlson D S and Brookes P C 1987 Measurement of soil microbial biomass provides and early indication of changes in total soil organic matter due to straw incorporation. Soil Biology and Biochemistry 19: 159-164.

Roper M M and Gupta V V S R 1995 Management practices and soil biota. Australian Journal of Soil Research 33: 321-339.

Ryan M H and Graham J H 2002 Is there a role for arbuscular mycorrhizal fungi in production agriculture? Plant and Soil 244: 263-271.

Shi Z, Wang K, Bailey J S, Jordan C and Higgins A H 2002 Temporal changes in the spatial distribution of some soil properties on a temperate grassland site. Soil Use and Management 18: 353-362.

Smith S E, Robson A D and Abbott L K 1993 The involvement of mycorrhizas of genetically-dependent efficiency of nutrient uptake and use. Plant and Soil 146: 169-179.

Smith J U, Smith P, Coleman K, Hargreaves P R and Macdonald A J 2002 Using dynamic simulation models and the 'Dot-to-Dot' method to determine the optimal sampling times in field trials. Soil Use and Management 18: 370-375.

Sojka R E and Upchurch D R 1999 Reservations regarding the soil quality concept. Soil Science Society of America Journal 63: 1039-1054.

Sojka R E, Upchurch D R and Borlaug N E 2003 Quality soil management or soil quality management: Performance versus semantics. Advances in Agronomy 79: 1-68.

Sorn-srivichai P, Syers J K, Tillman R W and Cornforth I S 1988 An evaluation of water extraction as a soil-testing procedure for phosphorus II. Factors affecting the amounts of water-extractable phosphorus in field soils. Fertilizer Research 15: 225-236.

Sparling G P 1997 Soil microbial biomass, activity and nutrient cycling as indicators of soil health. *In:* Biological indicators of soil health. C E Pankhurst, B M Doube and V V S R Gupta (eds.) pp. 97-119. CAB International. Wallingford, Oxon. UK.

Stenberg B 1999 Monitoring soil quality of arable land: Microbiological indicators. Acta Agriculturae Scandinavica 49: 1-24.

Stockdale E A, Lampkin N H, Hovi M, Keatinge R, Lennartsson E K M, Macdonald D W, Padel S, Tattersall F H, Wolfe M S and Watson C A 2001 Agronomic and environmental implications of organic farming systems. Advances in Agronomy 70: 261-327.

Stout J D, Gof K M and Rafter T A 1981 Chemistry and turnover of naturally occurring resistant organic compounds in soil. *In:* Soil Biochemistry Vol 5. E A Paul and J N Ladd (eds.) pp. 1-73. Marcel Dekker. New York.

Strong D T, Sale P W G and Helyar K R 1998 The influence of the soil matrix on nitrogen mineralisation and nitrification. I. Spatial variation and a hierarchy of soil properties. Australian Journal of Soil Research 36: 429-447.

Swift M J 1997 Biological management of soil fertility as a component of sustainable agriculture: Perspectives and prospects with particular reference to tropical regions. *In:* Soil ecology in sustainable agricultural systems. L Brussaard and R Ferrera-Cerrato (eds.) pp. 137-159. CRC Press. New York.

Tiedje J M, Cho J C, Murray A, Treves D, Xia B and Zhou J 2001 Soil teeming with life: New frontiers for soil science. *In:* Sustainable management of soil organic matter. R M Rees, B C Ball and C A Watson (eds.) pp. 393-412. CAB International. Wallingford, Oxon. UK.

Torsvik V, Goksoy J and Daae F L 1990 (a) High diversity in DNA of soil bacteria. Applied Environmental Microbiology 56: 782-787.

Torsvik V, Salte K, Sorheim R and Goksoyr J 1990 (b) Comparison of phenotypic diversity and DNA heterogeneity in a population of soil bacteria. Applied Environmental Microbiology 56: 776-781.

Turco R F, Kennedy A C and Jawson M D 1994 Microbial indicators of soil quality. *In:* Defining soil quality for a sustainable environment. J W Doran, D C Coleman, D F Bezdicek and B A Stewart (eds.) pp. 73-90. Soil Science Society of America Special Publication No 35. Madison, WI.

von Lützow M, Leifeld J, Kainz M, Kögel-Knabner I and Munch J C 2002 Indications for soil organic matter quality in soils under different management. Geoderma 105: 243-258.

Walker J and Reuter D J 1996 Key indicators to assess farm and catchment soil health. *In:* Indicators of catchment health: A technical perspective. J Walker and D J Reuter (eds.) pp. 21-33. CSIRO. Melbourne, Australia.

Zak J C, Willig M R, Moorhead D L and Wildman H G 1994 Functional diversity of microbial communities: A quantitative approach. Soil Biology and Biochemistry 26: 1101-1108.

Chapter 2

Impact of Fauna on Chemical Transformations in Soil

Graham H.R. Osler

Faculty of Natural and Agricultural Sciences, School of Earth and Geographical Sciences, The University of Western Australia, Crawley 6009, WA, Australia, and Agriculture and Agri-Food Canada, Lethbridge Research Centre, P.O. Box 3000, 5403-1 Avenue South, Lethbridge, Alberta, T1J 4B1, Canada.

1.   INTRODUCTION

Agricultural soils contain a multitude of animals. Tiny single celled protozoa to animals several orders of magnitude larger than protozoa thrive in agricultural soils. These communities exist irrespective of management practice or biogeographical location. Soil zoologists are faced with the challenge of understanding how this complex community impacts upon the processes vital for sustainable agricultural production. There is considerable evidence that soil fauna have large impacts on soil chemical transformations. There is also good understanding of the effects of management practices on soil faunal community structure. This should allow reasonable predictions of soil fauna responses to changes in management to be made. However, information on soil fauna effects on soil chemical transformations is not aligned with information on the effect of management practices on soil fauna. Consequently, few clear recommendations can be made to land managers on how they can exploit this resource to improve sustainability from a soil chemical fertility perspective.

17

L.K. Abbott & D.V. Murphy (eds) Soil Biological Fertility - A Key to Sustainable Land Use in Agriculture. 17-35.
© 2007 *Springer.*

This chapter examines the impact of fauna on chemical transformations using an ecological framework which considers functional groups, abundance, relative abundance, and species richness of the soil fauna. Functional groups are considered as broad categories of soil fauna (e.g. bacterial feeding nematodes), relative abundance as the relative proportions of individuals from different species or groups, and species richness as the number of species within a functional group. This framework is used to establish principles for interpreting the consequences of changes in soil faunal communities following changes in management practices. Beare (1997) concluded that it was important to determine how differences in soil food web structure can be used to predict the sustainability of agricultural practices. The framework adopted here is an attempt to address this issue. The chapter focuses on the impact of soil fauna on carbon (C) and nitrogen (N) mineralisation as these are two important elements in agriculture which have received a large amount of attention in soil fauna studies. These two elements are integrally linked such that C mineralisation rates are positively correlated with gross N immobilisation (Recous *et al.* 1999). Nitrogen mineralisation rates can also be dependent upon C:N ratios of the organic matter, although other factors, such as polyphenol and lignin content are also important (Heal *et al.* 1997). Other elements are obviously important for agricultural production and soil fauna are known to impact upon the way that they are cycled in soil (Seastedt 1984) but these interactions are not considered here.

This chapter subsequently addresses the effect of management practices on the soil fauna using the same framework in an attempt to draw out the principles established in the first section. Finally there is a discussion of how soil fauna may be more fully exploited to enhance soil chemical fertility and identifies some potential areas for future research are identified.

2.   THE ROLE OF SOIL FAUNA IN CARBON AND NITROGEN MINERALISATION

2.1   Functional Group Effects on Chemical Transformations

Food web models have identified protozoa (principally amoebae) and nematodes (bacteriovores and predators) as the greatest faunal contributors to soil N supply (de Ruiter *et al.* 1993). These groups have been estimated to contribute up to 40% of net N mineralisation (de Ruiter *et al.* 1993). Other groups such as mites and collembola were estimated to contribute less than 2% to net N mineralisation. Overall, soil fauna can account for 30-40% of net N mineralisation but in some situations, such as where nutrients are limited, this contribution can be much greater (Brussaard *et al.* 1996).

Many microcosm studies have investigated the effect of individual functional groups or increasing the number of functional groups on organic matter decomposition and subsequent soil N supply. These studies generally show that the addition of functional groups increases CO_2 respiration and mineral N production well above levels in microcosms which contain only bacteria and fungi. These effects are apparent for the smallest to largest soil animals (e.g. protozoa, nematodes, enchytraeids, collembola: Setälä *et al.* 1991, collembola and isopods: Teuben and Roelofsman 1990, protozoa to isopods: Coûteaux *et al.* 1991, termites: Ji and Brune 2001, earthworms and millipedes: Tian *et al.* 1995, isopods: Kautz and Topp 2000).

The impact of individual functional groups is complex and can vary depending upon resource quality, soil moisture, soil type and species examined. The impacts of functional groups on mass loss of organic matter, C losses and leaching of N has been found to be more pronounced when substrate quality is poor (ie. has high C:N ratio, Seastedt 1984, Coûteaux *et al.* 1991, Tian *et al.* 1995). The reason for this is unclear but may be due to stimulation of the microbial communities by the soil fauna (Tian *et al.* 1997). Similarly, the effects of microarthropods on NH_4^+-N production may change with soil moisture content (Sulkava *et al.* 1996). Isopods can increase CO_2 production and soil nutrient supply by 10 to 20% but their effect is much greater in sandy soils than in silt (Kautz and Topp 2000). However, the effects observed may be specific to species used in these microcosm studies. For example, results from a study to identify earthworm species suitable for introduction into Australian agricultural soils indicated that only a few species would increase plant growth in the field (Blakemore 1997).

Organic matter decomposition models, which include the role of the soil fauna, predict that soil fauna will only affect decomposition rates in environments where abiotic factors are non-limiting, such as in tropical rainforests and tropical savannas (Swift *et al.* 1979, Lavelle *et al.* 1993). The models therefore suggest that soil fauna will contribute little to decomposition in many areas where agriculture is practiced. However, a number of exclusion studies have demonstrated that soil microarthropods can have large impacts on decomposition in areas where soil moisture (Vossbrinck *et al.* 1979, Santos and Whitford 1981), or temperature is limiting (Douce and Crossley 1982). Indeed, a Tydeid mite may be a keystone species for decomposition of organic matter in a desert environment (Whitford 1996). On average, the presence of microarthropods increases decomposition rates by 23% across a range of environments (Seastedt 1984).

Only one study has systematically addressed this generalised model of decomposition (Heneghan *et al*. 1999). Microarthropods had much greater effects on decomposition of a single resource type at two tropical sites in comparison with a temperate site over a one-year period. Arthropods did not affect decomposition in the first 300 days of the study at the temperate site, whilst effects of arthropods were apparent at the tropical sites within the first month. However, decomposition rates at the temperate site were different between control and arthropod excluded litterbags on the last two sample dates which were greater than 300 days following litter placement. In fact, the arthropods at the temperate site began to show an effect on decomposition when 70% of the litter mass remained. Mass loss at one of the tropical sites was different within the first 50 days of the experiment and mass loss in the control bags at this time was approximately 80% (Heneghan *et al*. 1999). Similarly, delayed effects of arthropods on decomposition in temperate regions have also been shown in other studies (Vreeken-Buijs and Brussaard 1996, Blair *et al*. 1992). Therefore, in temperate ecosystems, there is the possibility that the effects of soil fauna on decomposition may only be apparent after a relatively long period of time and may only occur when the substrate has reached a particular state of decomposition. Indeed, it has been suggested that soil microarthropods do not affect decomposition rates during the leaching phase at the start of the decomposition process (Takeda 1995).

2.2   Abundance of Soil Fauna and their effect on Chemical Transformations

On a basic level, formulas used in food web models (e.g. Hunt *et al*. 1987) suggest that whenever abundance of a group increases, its contribution to N mineralisation increases. However, interactions between different groups within a food web will largely determine the extent of N mineralisation overall. For example, increased abundance of soil-dwelling earthworms can reduce numbers of surface-dwelling earthworms (Subler *et al*. 1997). The role of interactions on soil processes is discussed in Chapter 9.

Increased abundance of collembolans, isopods and millipedes has been shown to have a non-linear effect on CO_2 respiration (Hanlon and Anderson 1979, 1980). For example, microbial CO_2 respiration increased as the numbers of the collembolan *Folsomia candida* increased with 5 to 10 animals per microcosm, was unchanged with 15 animals per microcosm and decreased to levels below the control (no collembolans) with 20 collembolans per microcosm (Hanlon and Anderson 1979). Millipede abundance was also shown to have a non-linear effect on NH_4^+-N leached from microcosms (Anderson *et al.* 1983). In contrast to these studies, lower numbers of collembolans have been found to have the greatest impacts on N immobilisation (Mebes and Filser 1998). However, this effect was attributed to over-grazing by the collembolans in the early stages of their experiment, so that abundance of collembolans was very low when numbers were finally assessed (Mebes and Filser 1998). Others have found variable impacts of increased abundance of mites and collembolans depending upon the substrate used and duration of the study (Schulz and Scheu 1994, Kaneko *et al.* 1998).

How chemical transformations are affected by the abundance of soil fauna in the field is unclear. Direct correlation between microarthropod abundance and litter decomposition is unlikely, as their effects on these processes are indirect (Moore *et al.* 1988). In this light, there are contrasting results on the effect of higher microarthropod abundance on decomposition rates (House *et al.* 1987, Hendrix and Parmelee 1985). Reduction of nematode and microarthropod abundances in natural ecosystems through the application of biocides has been found to increase soil NH_4^+-N and NO_3^--N up to one year after applying the treatment (Ingham *et al.* 1989). This effect was apparent in all ecosystems studied but there were different rates of changes in inorganic N depending on the ecosystem.

The effect of greater earthworm abundance in agricultural fields may be dependent upon the sources of N. Blair *et al.* (1997) investigated the effect of changes in earthworm abundances on inorganic N supply in fields receiving three sources of N (inorganic, manure or cover crop additions). There was no effect of increased earthworm abundance (*Lumbricus terrestris*) in the manure and cover crop treatments. In the inorganic treatments, higher numbers of earthworms were associated with increased soil NO_3^--N at depths between 15 and 45 cm. This may have been caused by the effects of earthworms on soil porosity. The authors concluded that increased abundance of earthworms may increase NO_3^--N leaching to depth in inorganically fertilised fields (Blair *et al.* 1997). Subler *et al.* (1997) also found that increased abundance of *Lumbricus terrestris* may lead to losses of soil N, although in their study, increased earthworm abundance affected volumes of dissolved organic N (DON) in leachate rather than NO_3^--N.

2.3 Relative Abundance of Soil Fauna and their effect on Chemical Transformations

Few studies have examined how changes in the relative abundance of species within functional groups affect chemical transformations. The study by Heneghan and Bolger (1996) is an exception. They showed that CO_2 production and leaching of nutrients varied with microarthropod community structures, which had been altered through exposure to different toxins.

Similarly, mite diversity (measured using the Shannon index) is strongly correlated with gross N immobilisation rates in agricultural soils, despite the fact that these groups are not expected to contribute greatly to net N mineralisation rates (Table 1). In contrast, abundance and species richness of the mite communities were not related to gross N fluxes, although species richness was not fully determined at the study sites (Table 1). This is obviously only a correlation and does not reflect causation, but it demonstrates that soil faunal groups can strongly reflect soil chemical transformations that are the consequences of interactions within the soil food web.

Table 1 Correlation coefficients between measures of mite community structure and gross N fluxes. * significantly different from zero at P<0.01. (Osler *et al.* unpublished data).

Community attribute	N mineralisation	N immobilisation
Abundance	-0.075	-0.080
Species richness	0.073	0.664
Species diversity (Shannon index)	0.356	0.956*

2.4   Species Richness of Soil Fauna and the effect on Chemical Transformations

The role of species richness in ecosystem processes is presently one of the most debated topics in ecology, although there have been relatively few studies addressing this question in soil ecosystems (Giller *et al.* 1997, Swift *et al.* 1998). Andrén *et al.* (1995) observed no correlation between mass loss of organic matter and the community structure of soil faunal groups (measured using diversity indices), and concluded that this implied species redundancy in these communities. However, they noted that there may have been a successional effect of the fauna on decomposition, and therefore a role for species diversity in the decomposition process. Only one study has identified a successional pattern of soil microarthropods during decomposition of leaf litter (Santos and Whitford 1981), although a number of studies have shown a similar pattern of change to that described by them (e.g. Vreeken-Buijs and Brussaard 1996, Osler *et al.* 2000).

Studies which rigorously addressed the species richness question have generally supported an idiosyncratic response hypothesis (Mikola and Setälä 1998, Laakso and Setälä 1999). That is, different levels of species richness changed CO_2 and NH_4^+-N production but the direction of the change was unpredictable and required knowledge of species characteristics and interactions. Nevertheless, levels of species richness significantly alter CO_2 production (Mikola and Setälä 1998) and there can be large differences in

NH_4^+-N production between treatments with different levels of species richness (up to 50%: Laakso and Setälä 1999).

Amounts of soil inorganic N can vary between single and multiple collembolan species assemblages. Mebes and Filser (1998) found that in treatments with 5 collembolan species there was net mineralisation of inorganic N whilst in treatments containing only one collembolan species there was net N immobilisation. This effect was found for all of the species included in the experiment. The difference in N effects in the multiple compared with the single species communities was attributed to over exploitation of a single resource in the single species treatments (Mebes and Filser 1998).

2.5 Summary of Faunal Impacts on Chemical Transformations

The discussion above demonstrates that a large array of soil fauna have significant impacts on soil processes and chemical transformations. Mass loss of organic matter, CO_2 and NH_4^+-N production, and NO_3^--N leaching can all be affected by the presence of different functional groups, the abundance of a single species, and the relative abundance of animals. There is considerably more information on some aspects of the effects of soil fauna on chemical transformations; the contribution of soil faunal functional groups to chemical transformations has been well documented, whilst there are few studies on the role of abundance, relative abundance and species richness.

The studies discussed here indicate that the impact of soil fauna on soil organic matter transformations will change with factors such as resource quality, soil moisture and the duration of the study. The few studies that have contrasted animal impacts with differences in these factors indicate that soil fauna have greatest effects when resource quality is poor, although the reason for this is unclear.

3. MANAGEMENT IMPACTS ON SOIL FAUNA

3.1 Management Impacts on Functional Richness

A limited number of studies have demonstrated an increase in functional richness of soil fauna with a change in agricultural practice. Wardle (1995) conducted a meta-analysis of the effects of tillage on the soil biota, and noted that the impact of tillage depended upon the frequency and intensity of the tillage regime. With this in mind, new functional groups can enter reduced

tillage systems. For example, Didden *et al.* (1994) found earthworms in their
no-till treatment when none occurred in the conventional tillage treatment.
In their food web model of these two systems they estimated that the
contribution of soil fauna to net N mineralisation was 4.3 and 39.4% in the
conventional and no-till treatments, respectively. This dramatic change was
largely due to the presence of earthworms in the no-till treatment (Didden *et
al.* 1994). The change in the contribution of the fauna between the two
systems did not lead to increased N mineralisation in the no-till treatment,
although N dynamics over time and by depth was different between the two
systems (de Ruiter *et al.* 1994). Other studies (e.g. Franchini and Rockett
1996, showing the appearance of a macrophytophagous oribatid mite) found
greater numbers of functional groups following changes in tillage regimes.

3.2   Management Impacts on the Abundance of Soil Fauna

In contrast to changes in functional richness, many management
practices have large impacts on the abundance of different soil fauna groups.
Plant species have different effects on the abundance of individual soil fauna
species or communities. For example, abundance of termites and
earthworms are differentially affected by mulches whilst abundance of
millipedes remained unchanged with the same treatments (Tian *et al.* 1993).
Yeates *et al.* (1999) found that nematode numbers were reduced with the
application of pine sawdust as mulch, whilst in the same treatment, the
numbers of macroarthropods (beetles and spiders) were increased (Wardle *et
al.* 1999). Further, the effect of the mulch treatment was greater than the
effect of cultivation (Wardle *et al.* 1999). In nematode communities, the
abundance of bacteriovores may respond to litter quality, whilst quantity of
litter alone may be more important for fungivores (Mikola *et al.* 2001). The
more favoured plant species for increasing the abundance of microarthropods
is a complex interaction between residue quality and residence time of the
organic matter (Badejo *et al.* 1995).

In terms of changes in abundance of soil fauna following changes in
tillage, Wardle's (1995) meta-analysis showed that, in general, tillage had the
greatest impact on the abundance of larger animals such as beetles and
earthworms (Figure 1). However, the abundance of functional groups of
smaller soil fauna can also change dramatically with tillage. In experiments
at Horseshoe Bend, USA, the abundance of bactivorous nematodes was
reduced in no-tillage systems whilst the abundance of fungivorous
nematodes increased (Parmelee and Alston 1986). In this experiment,
abundance of a number of functional groups altered with changes in tillage
(beetles, earthworms, spiders, mites and collembolans; Hendrix *et al.* 1986)

but there was no apparent increase in functional richness. Several authors have demonstrated that the abundance of soil animals can remain high in no-till systems when dry conditions are experienced due to greater moisture holding capacity of the soils (e.g. Elliott *et al.* 1984, Perdue and Crossley 1989), whilst in conventionally cultivated fields weeds can be an important food supply for soil fauna where there are few other organic matter resources (Garrett *et al.* 2001).

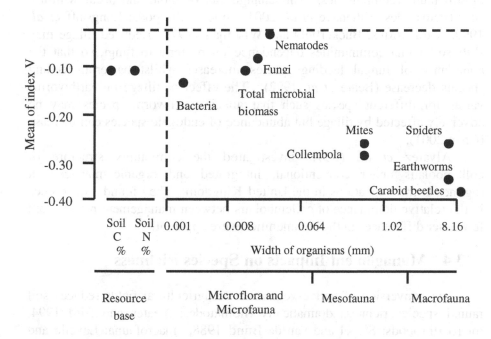

Figure 1 The impact of tillage on the abundance of different fauna groups from a meta-analysis by Wardle (1995). The y-axis represents a decrease in abundance of groups with tillage on a relative scale to allow comparisons between studies. The more negative the value the greater the reduction in abundance of a group. Reproduced with permission.

Livestock grazing has been shown to increase the abundance of different groups of fauna, but if grazing has a large impact on soil pore volume, abundance can be reduced (Bardgett and Cook 1998). Similarly, the use of fertiliser can increase the abundance of groups such as nematodes, rotifers and tardigrades (McIntosh *et al.* 1999). However, if fertilisers induce large reductions in soil pH there can be negative effects on the abundance of soil fauna (e.g. Hansen and Engelstad 1999).

3.3   Management Impacts on Relative Abundance

Plant species alter the relative abundance of different faunal groups, but do not appear to affect functional and species richness (eg. Hansen 1999, Osler and Beattie 2001). These changes in soil fauna structure may lead to altered decomposition rates (Hansen 1999).

Changing tillage practice can considerably alter the relative abundance of soil faunal communities. This change can be rapid and occur within a year (nematodes: Villenave *et al*. 2001, microarthropods: Longstaff *et al*. 1999, earthworms: Buckerfield and Wiseman 1997). Reduced tillage may shift soil fauna communities dependence on bacteria to fungi, so that the abundance of fungal feeding species increases whilst bacterial feeding species decrease (Beare *et al*. 1992). The effect of tillage on earthworms varies for different species such that anecic earthworm species may be adversely affected by tillage but abundance of endogeic species can increase (Chan 2001).

Alvarez *et al*. (2001) investigated the community structure of collembolans under conventional, integrated and organic management regimes at three locations in the United Kingdom. They found a difference in the relative dominance of collembolans between management regimes but few other differences in the communities were apparent.

3.4   Management Impacts on Species Richness

The conversion of native vegetation to agricultural fields reduces soil faunal species richness dramatically (nematodes: Yeates and Bird 1994, microarthropods: Siepel and van de Bund 1988, macrofauna: Lavelle and Pashanasi 1989, Dangerfield 1990). Changing land use also favours particular groups of animals, so, for example, abundance of prostigmatid mites may increase under agriculture compared with native vegetation whilst abundance of oribatids may decline (Crossley *et al*. 1992). Whether species in surrounding native vegetation return to agricultural land following changes in management practice depends upon the dispersal modes of the animals and the landscape matrix of the surrounding area (Giller *et al*. 1997). Practices such as tillage can reduce species richness dramatically (e.g. Neave and Fox 1998), however, this is not always the case as other factors are important: species richness of beetles can be greater in conventionally tilled plots compared with reduced tillage plots at particular times, due to the presence of weeds (and therefore greater food resources) in conventionally tilled plots (Wardle *et al*. 1999). Similarly, pesticides can have variable impacts on the soil fauna community depending upon their toxicity to species or groups and whether they increase or decrease soil organic matter (Wardle 1995, Neher 1999).

3.5 Summary of Management Impacts on the Soil Fauna

In general, the effects of management practices on the soil fauna are well understood. Reducing tillage tends to increase the abundance of fauna, change the relative abundance of faunal groups and occasionally alters functional or species richness. Plant species can alter abundance and relative abundance of soil fauna but they generally do not change functional or species richness. Practices that retain plant cover, such as retention of weeds and use of pastures, tend to increase the abundance of animals. Crossley *et al.* (1992) summarised the effects of agricultural practices on the soil biota. They stated that the use of fertilisers and polycultures were likely to lead to increased species diversity and population densities, whilst the use of pesticides, tillage and monocultures were likely to lead to reduced species diversity and population increases only in some species. Such effects are dependent upon a number of factors. For example, in Austria, few differences in the soil biota between conventional and ecofarming were apparent in areas with sufficient rainfall and mixed farming practices, but biological activity was reduced in conventionally managed fields in more arid areas (Foissner 1992). Similarly, organic management regimes in grassland can affect groups of soil fauna, but not to the same extent as in arable fields (Yeates *et al.* 1997).

Understanding the effects of changes in abundance and relative abundance of soil faunal groups on chemical transformations is essential for identifying the potential contributions soil fauna can make to nutrient cycling in agroecosystems; many of the studies of management impacts on soil fauna show that one of the main responses is a change in abundance. The general trend in abundance studies is decreased CO_2 and N mineralisation with high faunal abundance. Therefore, where management practices increase faunal abundance, these changes may be observed. In general, reducing tillage tends to increase soil faunal abundance, decrease decomposition rate and increase nutrient immobilisation (Hendrix *et al.* 1986). However, tillage alters many factors, such as the distribution of organic matter and soil structure (Ghuman and Sur 2001, Kushwaha *et al.* 2001) and changes in nutrient transformations can not be attributed to the soil fauna alone. Further, the work by Blair *et al.* (1997) on changes in earthworm abundance shows that there can be complex interactions between an increase in the abundance of a group of animals and the agricultural management regime.

The two sections above demonstrate a discrepancy between soil biologists' knowledge of the role of soil fauna in chemical transformations and the impact of land management practices on the soil fauna. Much work has been conducted on the effects of different functional groups on chemical transformations, but only a few studies have shown a change in functional or species richness with changes in management practices. Further, the majority of the research to date on functional richness has used microcosm studies where the aim was to determine how different groups interact and affect chemical transformations. Agricultural fields already contain the trophic groups that are commonly used in these microcosm experiments; it would be difficult to find any agricultural soils that did not contain protozoa, nematodes, mites, collembolans and many other groups. Therefore, one of the challenges for soil ecologists applying their work to agricultural systems is to examine the consequences of additions of functional groups to already complex communities and to understand how different levels of abundance of groups of soil fauna influence soil processes (Blair *et al.* 1997, Bardgett and Cook 1998).

4.   IMPROVING SOIL CHEMICAL FERTILITY WITH SOIL FAUNA

Recommendations on how soil fauna might be managed to improve soil chemical fertility are dependant upon the aims of the land manager. In terms of soil N, the aims of agriculturalists might be optimal supply of nutrient, synchronous supply of nutrient with crop demands, and minimal losses of nutrients through leaching (Beare 1997, Neher 1999). Soil fauna could contribute to all of these aims, although, as outlined above, soil biologists need to define the community structures necessary for these aims to be achieved in the field. Thus, soil biologists are in the same situation as other agricultural scientists, such as agronomists, who are beginning to investigate techniques that will reduce the need for high levels of inputs (e.g. selecting suitable green manures). The objectives for soil C are less apparent than for N because C is not a plant nutrient. However, conserving and or increasing soil organic matter is desirable, as is reducing stubble loads to reduce disease risk (Jarvis *et al.* 1996, Kumar and Goh 2000). The impact of soil fauna on organic matter decomposition rates suggests they will contribute to this process but that their effects may not always be immediately apparent. Further, soil fauna have greatest impacts on the poorest quality litter.

Land managers can use either direct or indirect measures to alter soil faunal community structure and therefore manipulate soil chemical transformations. Indirect techniques include: reducing tillage, using organic amendments, and maintaining plant growth. These techniques tend to increase the abundance of soil fauna and alter the relative abundance of groups. This increased abundance appears to lead to greater immobilisation of nutrients. In terms of building and maintaining soil fertility this is a desirable outcome. Presently there are very few direct ways to increase soil fauna contributions to chemical transformations. However, there is much scope to explore the potential of soil biology to improve soil fertility. Such techniques could include selecting plant species based on their effects on the soil biota and the re-introduction of species of fauna excluded through past agricultural practices.

The non-linear response of chemical transformations with changes in faunal abundance (Hanlon and Anderson 1979, 1980) suggests that there may be an optimal abundance of soil faunal groups for chemical transformations. When examining possible new plant species or varieties as crops or green manures, could their effect on the soil biota be used as one of the selection criteria? This would require soil biologists to stipulate the types of community structures most favourable for the aims of agriculturalists. Such prescriptions are likely to be difficult to construct and to be site or regionally specific.

Species richness may impact on ecosystem processes in numerous ways, such as resilience and resistance to perturbations (Pimm 1991). Although neither Laakso and Setälä (1999) or Mikola and Setälä (1998) found evidence that greater species richness increased soil nutrient supply they concluded that species richness may play other important roles in the soil. Ecosystem traits such as resilience are likely to be useful to agriculturalists who manage systems that are constantly subjected to disturbances (e.g. drought and disease). The role of soil fauna species richness in chemical transformations is poorly understood at present but developing this area of research could contribute to establishing more sustainable agricultural practices.

Few studies have found increased species richness following changes in management practice and yet there are likely to be large numbers of species outside agricultural areas which have the potential to colonise agricultural fields. This suggests that either the return of many groups is slow or the altered agricultural environments do not provide suitable habitats for these species. In this light, it is well known that recovery of soil fauna populations following different disturbances can take decades (Curry and Good 1992, Greenslade and Major 1993, Webb 1994). There have been few attempts to re-introduce species where management practices have been altered (e.g. when reduced tillage systems are implemented) although there have been many calls for this to be further investigated (e.g. Abbott *et al*. 1979, Baker 1998). The work by Blakemore (1997) on earthworms in Australia provides an example of how such a process could be conducted. The large differences in species richness of soil faunal groups between agricultural and native vegetation soils suggests that there are many groups which have the potential to be exploited. These species may represent new functional groups or augment the species richness of functional groups already present. A number of important soil fauna groups with different functions could be investigated (Andrén *et al.* 1999). One such group of species are those which are active in the non-growing season. For example, in the Western Australian wheatbelt, which has a Mediterranean type climate, large amounts of N can be mineralised during infrequent rainfall events over the typically dry summer period (Murphy *et al.* 1998). This rapid mineralisation can contribute to soil acidification and loss of soil N. There may be summer active soil fauna groups which could alleviate this problem through immobilisation of nutrients, as soil fauna in natural ecosystems can respond rapidly to episodic changes in soil moisture (Whitford *et al.* 1981, Osler *et al.* 2001).

5.   CONCLUSION

Soil fauna have a large impact on the chemical transformations which are vital for sustainable agricultural production. Presently this resource is not directly managed and consequently under-exploited. With better knowledge of the effects of some key community parameters on chemical transformations, such as determining optimal abundance of fauna, soil biologists will be able to make a greater contribution to the design of more sustainable agricultural practices.

6.   REFERENCES

Abbott I, Parker C A and Sills I D 1979 Changes in the abundance of large soil animals and physical properties of soils following cultivation. Australian Journal of Soil Research 17: 343-353.

Alvarez T, Frampton G K and Goulson D 2001 Epigeic collembola in winter wheat under organic, integrated and conventional farm management regimes. Agriculture, Ecosystems and Environment 83: 95-110.

Anderson J M, Ineson P and Huish S A 1983 Nitrogen and cation mobilisation by soil fauna feeding on leaf litter and soil organic matter from deciduous woodlands. Soil Biology and Biochemistry 15: 463-467.

Andrén O, Brussaard L and Clarholm M 1999 Soil organism influence on ecosystem-level processes - bypassing the ecological hierarchy? Applied Soil Ecology 11: 177-188.

Andrén O, Bengtsson J and Clarholm M 1995 Biodiversity and species redundancy among litter decomposers. *In:* The Significance and Regulation of Soil Biodiversity. H P Collins, G P Robertson and M J Klug (eds) pp. 141-151. Kluwer Academic Publishers. The Netherlands.

Badejo M A, Tian G and Brussaard L 1995 Effects of various mulches on soil microarthropods under a maize crop. Biology and Fertility of Soils 20: 294-298.

Baker G H 1998 Recognising and responding to the influences of agriculture and other land-use practices on soil fauna in Australia. Applied Soil Ecology 9: 303-310.

Bardgett R D and Cook R 1998 Functional aspects of soil animal diversity in agricultural grasslands. Applied Soil Ecology 10: 263-276.

Beare M 1997 Fungal and bacterial pathways of organic matter decomposition and nitrogen mineralisation in arable soils. *In:* Soil Ecology in Sustainable Agricultural Systems. L Brussaard and R Ferrera-Cerrato (eds.) pp. 37-70. CRC Press. USA.

Beare M H, Parmelee R W, Hendrix P F, Cheng W, Coleman D C and Crossley D A 1992 Microbial and faunal interactions and effects on litter nitrogen and decomposition in agroecosystems. Ecological Monographs 62: 569-591.

Blair J M, Parmelee R W, Allen M F, McCartney D A and Stinner B R 1997 Changes in soil N pools in response to earthworm population manipulations in agroecosystems with different N sources. Soil Biology and Biochemistry 29: 361-367.

Blair J M, Crossley D A and Callaham L C 1992 Effects of litter quality and microarthropods on N dynamics and retention of exogenous ^{15}N in decomposing litter. Biology and Fertility of Soils 12: 241-252.

Blakemore R J 1997 Agronomic potential of earthworms in brigalow soils of south-east Queensland. Soil Biology and Biochemistry 29: 603-608.

Brussaard L, Bakker J P and Olff H 1996 Biodiversity of soil biota and plants in abandoned arable fields and grasslands under restoration management. Biodiversity and Conservation 5: 211-221.

Buckerfield J C and Wiseman D M 1997 Earthworm populations recover after potato cropping. Soil Biology and Biochemistry 29: 609-612.

Chan K Y 2001 An overview of some tillage impacts on earthworm population abundance and diversity - implications for functioning in soils. Soil and Tillage Research 57: 179-191.

Coûteaux M, Mousseau M, Celerier M and Bottner P 1991 Increased atmospheric CO_2 and litter quality: decomposition of sweet chestnut leaf litter with animal food webs of different complexities. Oikos 61: 54-64.

Crossley D A, Mueller B R and Perdue J C 1992 Biodiversity of microarthropods in agricultural soils: relations to processes. Agriculture, Ecosystems and Environment 40: 37-46.

Curry J P and Good J A 1992 Soil faunal degradation and restoration. Advances in Soil Science 17: 171-215.

Dangerfield J M 1990 Abundance, biomass and diversity of soil macrofauna in savanna woodland and associated managed habitats. Pedobiologia 34: 141-150.

de Ruiter P C, Bloem J, Bouwman L A, Didden W A M, Hoenderboom G H J, Lebbink G, Marinissen J C Y, de Vos J A, Vreeken-Buijs M J, Zwart K B and Brussaard L 1994 Simulation of dynamics in nitrogen mineralisation in the belowground food webs of two arable farming systems. Agriculture, Ecosystems and Environment 51: 199-208.

de Ruiter P C, Moore J C, Zwart K B, Bouwman L A, Hassink J, Bloem J, de Vos J A, Marinissen J C Y, Didden W A M, Lebbink G and Brussaard L 1993 Simulation of nitrogen mineralisation in the belowground food webs of two winter wheat fields. Journal of Applied Ecology 30: 95-106.

Didden W A M, Marinissen J C Y, Vreeken-Buijs M J, Burgers S L G E, de Fluiter R, Geurs M and Brussaard L 1994 Soil meso- and macrofauna in two agricultural systems: factors affecting population dynamics and evaluation of their role in carbon and nitrogen dynamics. Agriculture, Ecosystems and Environment 51: 171-186.

Douce G K and Crossley D A 1982 The effect of soil fauna on litter mass loss and nutrient loss dynamics in arctic tundra at Barrow, Alaska. Ecology 63: 523-537.

Elliott E T, Horton K, Moore J C, Coleman D C and Cole C V 1984 Mineralisation dynamics in fallow dryland wheat plots, Colorado. Plant and Soil 76: 149-155.

Foissner W 1992 Comparative studies on the soil life in ecofarmed and conventionally farmed fields and grasslands of Austria. Agriculture, Ecosystems and Environment 40: 207-218.

Franchini P and Rockett C L 1996 Oribatid mites as "indicator" species for estimating the environmental impact of conventional and conservation tillage practices. Pedobiologia 40: 217-225.

Garrett C J, Crossley D A, Coleman D C, Hendrix P F, Kisselle K W and Potter R L 2001 Impact of the rhizosphere on soil microarthropods in agroecosystems on the Georgia piedmont. Applied Soil Ecology 16: 141-148.

Ghuman B S and Sur H S 2001 Tillage and residue management effects on soil properties and yields of rainfed maize and wheat in a subhumid subtropical climate. Soil and Tillage Research 58: 1-10.

Giller K E, Beare M H, Lavelle P, Izac A M N and Swift M J 1997 Agricultural intensification, soil biodiversity and agroecosystem function. Applied Soil Ecology 6: 3-16.

Greenslade P and Majer J D 1993 Recolonisation by collembola of rehabilitated bauxite mines in Western Australia. Australian Journal of Ecology 18: 385-394.

Hanlon R D G and Anderson J M 1980 Influence of macroarthropod feeding activities on microflora in decomposing oak leaves. Soil Biology and Biochemistry 12: 255-261.

Hanlon R D G and Anderson J M 1979 The effects of collembola grazing on microbial activity in decomposing leaf litter. Oecologia 38: 93-99.

Hansen R A 1999 Red oak litter promotes a microarthropod functional group that accelerates decomposition. Plant and Soil 209: 37-45.

Hansen S and Engelstad F 1999 Earthworm populations in a cool and wet district as affected by tractor traffic and fertilisation. Applied Soil Ecology 13: 237-250.

Heal O W, Anderson J M and Swift M J 1997 Plant litter quality and decomposition: an historical overview. *In:* Driven by Nature: Plant Litter Quality and Decomposition. G Cadisch and G E Giller (eds.) pp. 3-30. CAB International, Wallingford, Oxon. UK.

Hendrix P F, Parmelee R W, Crossley D A, Coleman D C, Odum E P and Groffman P M 1986 Detritus food webs in conventional and no-tillage agroecosystems. Bioscience 36: 374-380.

Hendrix P F and Parmelee R W 1985 Decomposition, nutrient loss and microarthropod densities in herbicide-treated grass litter in a Georgia piedmont agroecosystem. Soil Biology and Biochemistry 17: 421-428.

Heneghan L, Coleman D C, Zou X, Crossley D A and Haines B L 1999 Soil microarthropod contributions to decomposition dynamics: tropical-temperate comparisons of a single substrate. Ecology 80: 1873-1882.

Heneghan L and Bolger T 1996 Effects of components of 'acid rain' on the contribution of soil microarthropods to ecosystem functioning. Journal of Applied Ecology 33: 1329-1344.

House G J, Worsham A D, Sheets T J and Stinner R E 1987 Herbicide effects on soil arthropod dynamics and wheat straw decomposition in a North Carolina no-tillage agroecosystem. Biology and Fertility of Soils 4: 109-114.

Hunt H W, Coleman D C, Ingham E R, Ingham R E, Elliot E T, Moore J C, Rose S L, Reid C P P and Morley C R 1987 The detrital food web in a shortgrass prairie. Biology and Fertility of Soils 3: 57-68.

Ingham E R, Coleman D C and Moore J C 1989 An analysis of food-web structure and function in a shortgrass prairie, a mountain meadow, and a lodgepole pine forest. Biology and Fertility of Soils 8: 29-37.

Jarvis S C, Stockdale E A, Shepherd M A and Powlson D S 1996 Nitrogen mineralisation in temperate agricultural soils: processes and measurement. Advances in Agronomy 57: 187-235.

Ji R and Brune A 2001 Transformation and mineralization of ^{14}C-labeled cellulose, peptidoglycan, and protein by the soil-feeding termite *Cubitermes orthognathus*. Biology and Fertility of Soils 33: 166-174.

Kaneko N, McLean M A and Parkinson D 1998 Do mites and Collembola affect pine litter fungal biomass and microbial respiration? Applied Soil Ecology 9: 209-213.

Kautz G and Topp W 2000 Acquisition of microbial communities and enhanced availability of soil nutrients by the isopod *Porcellio scaber* (Latr.) (Isopoda: Oniscidea). Biology and Fertility of Soils 31: 102-107.

Kumar K and Goh K M 2000 Crop residues and management practices: effects on soil quality, soil nitrogen dynamics, crop yield, and nitrogen recovery. Advances in Agronomy 68: 197-319.

Kushwaha C P, Tripathi S K and Singh K P 2001 Soil organic matter and water-stable aggregates under different tillage and residue conditions in a tropical dryland agroecosystem. Applied Soil Ecology 16: 229-241.

Laakso J and Setälä H 1999 Sensitivity of primary production to changes in the architecture of belowground food webs. Oikos 87: 57-64.

Lavelle P, Blanchart E, Martin A, Martin S, Spain A, Toutain F, Barios I and Schaefer R 1993 A hierarchical model for decomposition in terrestrial ecosystems: application to soils of the humid tropics. Biotropica 25: 130-150.

Lavelle P and Pashanasi B 1989 Soil macrofauna and land management in Peruvian Amazonia (Yurimagus, Loreto). Pedobiologia 33: 283-291.

Longstaff B C, Greenslade P J M, Colloff M, Reid I, Hart P and Packer I 1999 Managing soils in agriculture, the impact of soil tillage practices on soil fauna. RIRDC Publication No. 99/18. Canberra, Australia.

McIntosh P D, Gibson R S, Saggar S, Yeates G W and McGimpsey P 1999. Effect of contrasting farm management on vegetation and biochemical, chemical, and biological condition of moist steepland soils of the South Island high country, New Zealand. Australian Journal of Soil Research 37: 847-865.

Mebes K H and Filser J 1998 Does the species composition of Collembola affect nitrogen turnover? Applied Soil Ecology 9: 241-247.

Mikola J and Setälä H 1998 Relating species diversity to ecosystem functioning: mechanistic backgrounds and experimental approach with a decomposer food web. Oikos 83: 180-194.

Mikola J, Yeates G W, Wardle D A, Barker G M and Bonner K I 2001. Response of soil food-web structure to defoliation of different plant species combinations in an experimental grassland community. Soil Biology and Biochemistry 33: 205-214.

Moore J C, Walter D E and Hunt H W 1988 Arthropod regulation of micro- and mesobiota in below-ground detrital food webs. Annual Review of Entomology 33: 419-439.

Murphy D V, Sparling G P, Fillery I R P, McNeill A M and Braunberger P 1998 Mineralisation of soil organic nitrogen and microbial respiration after simulated summer rainfall events in an agricultural soil. Australian Journal of Soil Research 36: 231-246.

Neave P and Fox C A 1998 Response of soil invertebrates to reduced tillage systems established on a clay loam soil. Applied Soil Ecology 9: 423-428.

Neher D A 1999 Soil community composition and ecosystem processes: comparing agricultural ecosystems with natural ecosystems. Agroforestry Systems 45: 159-185.

Osler G H R and Beattie A J 2001 Contribution of oribatid and mesostigmatid soil mites in ecologically based estimates of global species richness. Austral Ecology 26: 70-79.

Osler G H R, Westhorpe D and Oliver I 2001 The short-term effects of endosulfan discharges on eucalypt floodplain soil microarthropods. Applied Soil Ecology 16: 263-273.

Osler G H R., van Vliet P C J, Gauci C S and Abbott L K 2000 Changes in free living soil nematode and micro-arthropod communities under a canola-wheat-lupin rotation in Western Australia. Australian Journal of Soil Research 38: 47-59.

Parmelee R W and Alston D G 1986 Nematode trophic structure in conventional and no-tillage agroecosystems. Journal of Nematology 18: 403-407.

Perdue J C and Crossley D A 1989 Seasonal abundance of soil mites (Acari) in experimental agroecosystems: effect of drought in no-tillage and conventional tillage. Soil and Tillage Research 15: 117-124.

Pimm S L 1991 The balance of nature? Ecological issues in the conservation of species and communities. University of Chicago Press. Chicago.

Recous S, Aita C and Mary B 1999 In situ changes in gross N transformations in bare soil after addition of straw. Soil Biology and Biochemistry 31: 119-133.

Santos P F and Whitford W G 1981 The effects of microarthropods on litter decomposition in a Chihauhuan desert ecosystem. Ecology 62: 654-663.

Schulz E and Scheu S 1994 Oribatid mite mediated changes in litter decomposition: model experiments with ^{14}C-labelled holocellulose. Pedobiologia 38: 344-352.

Seastedt T R 1984 The role of microarthropods in decomposition and mineralisation processes. Annual Review of Entomology 29: 25-46.

Setälä H, Tyynismaa M, Martikainen E and Huhta V 1991 Mineralisation of C, N and P in relation to decomposer community structure in coniferous forest soil. Pedobiologia 35: 285-296.

Siepel H and van de Bund C F 1988 The influence of management practises on the microarthropod community of grassland. Pedobiologia 31: 339-354.

Subler S, Baranski C M and Edwards C A 1997 Earthworm additions increased short-term nitrogen availability and leaching in two grain-crop agroecosystems. Soil Biology and Biochemistry 29: 413-421.

Sulkava P, Huhta V and Laakso J 1996 Impact of soil fauna structure on decomposition and N-mineralisation in relation to temperature and moisture in forest soil. Pedobiologia 40: 505-513.

Swift M J, Andrén O, Brussaard L, Briones M, Coûteaux M M, Ekschmitt K, Kjoller A, Loiseau P and Smith P 1998 Global change, soil biodiversity, and nitrogen cycling in terrestrial ecosystems: three case studies. Global Change Biology 4: 729-743.

Swift M J, Heal O W and Anderson J M 1979 Decomposition in terrestrial ecosystems. Blackwell Scientific. Oxford, UK.

Takeda H 1995 A 5 year study of litter decomposition processes in a *Chamaecyparis obtusa* Endl. forest. Ecological Research 10: 95-104.

Teuben A and Roelofsma T A P J 1990 Dynamic interactions between functional groups of soil arthropods and microorganisms during decomposition of coniferous litter in microcosm experiments. Biology and Fertility of Soils 9: 145-151.

Tian G, Brussaard L, Kang B T and Swift M J 1997 Soil fauna-mediated decomposition of plant residues under constrained environmental and residue quality conditions. *In:* Driven by Nature: Plant Litter Quality and Decomposition. G Cadisch and G E Giller (eds.) pp. 125-134. CAB International. Wallingford, Oxon, UK.

Tian G, Brussaard L and Kang B T 1995 Breakdown of plant residues with contrasting chemical compositions under humid tropical conditions: effects of earthworms and millipedes. Soil Biology and Biochemistry 27: 277-280.

Tian G, Brussaard L and Kang B T 1993 Biological effects of plant residues with contrasting chemical compositions under humid tropical conditions: effects on soil fauna. Soil Biology and Biochemistry 25: 731-737.

Villenave C, Bongers T, Ekschmitt K, Djibal D and Chotte J L 2001 Changes in nematode communities following cultivation of soils after fallow periods of different length. Applied Soil Ecology 17: 43-52.

Vossbrinck C R, Coleman D C and Wooley T A 1979 Abiotic and biotic factors in litter decomposition in a semiarid grassland. Ecology 60: 265-271.

Vreeken-Buijs M J and Brussaard L 1996 Soil mesofauna dynamics, wheat residue decomposition and nitrogen mineralisation in buried litter bags. Biology and Fertility of Soils 23: 374-381.

Wardle D A 1995 Impacts of disturbance on detritus food webs in agro-ecosystems of contrasting tillage and weed management practices. Advances in Ecological Research 26: 105-185.

Wardle D A, Nicholson K S, Bonner K I and Yeates G W 1999 Effects of agricultural intensification on soil-associated arthropod population dynamics, community structure, diversity and temporal variability over a seven-year period. Soil Biology and Biochemistry 31: 1691-1706.

Webb N R 1994 Post-fire succession of cryptostigmatic mites (Acari, Cryptostigmata) in *Calluna*-heathland soil. Pedobiologia 38: 138-145.

Whitford W G 1996 The importance of the biodiversity of soil biota in arid ecosystems. Biodiversity and Conservation 5: 185-195.

Whitford W G, Freckman D W, Elkins N Z, Parker L W, Parmelee R, Phillips J and Tucker S 1981 Diurnal migration and responses to simulated rainfall in desert soil microarthropods and nematodes. Soil Biology and Biochemistry 13: 417-425.

Yeates G W, Wardle D A and Watson R N 1999 Responses of soil nematode populations, community structure, diversity and temporal variability to agricultural intensification over a seven-year period. Soil Biology and Biochemistry 31: 1721-1733.

Yeates G W, Bardgett R D, Cook R, Hobbs P J, Bowling P J and Potter J F 1997 Faunal and microbial diversity in three Welsh grassland soils under conventional and organic management regimes. Journal of Applied Ecology 34: 453-470.

Yeates G W and Bird F 1994 Some observations on the influence of agricultural practices on the nematode faunae of some South Australian soils. Fundamental and Applied Nematology 17: 133-145.

Chapter 3

Impact of Microorganisms on Chemical Transformations in Soil

Daniel V. Murphy[1], Elizabeth A. Stockdale[2], Philip C. Brookes[2] and Keith W.T. Goulding[2]

[1]School of Earth and Geographical Sciences, Faculty of Natural and Agricultural Sciences, The University of Western Australia, Crawley, 6009, WA, Australia.
[2] Agriculture and Environment Division, Rothamsted Research, Harpenden, Hertfordshire, AL5 2JQ, United Kingdom.

1. INTRODUCTION

Microorganisms (e.g. bacteria, fungi, actinomycetes, microalgae) play a key role in organic matter decomposition, nutrient cycling and other chemical transformations in soil. In fact general measurements of microbial activity in soil are synonymous with the breakdown of organic matter. Decomposition of organic matter is usually controlled by heterotrophic microorganisms and leads to the release and cycling of nutrients (especially nitrogen (N), sulphur (S) and phosphorus (P)). Microorganisms also immobilise significant amounts of carbon (C) and other nutrients within their cells. The total mass of living microorganisms (the microbial biomass) therefore has a central role as source, sink and regulator of the transformations of energy and nutrients in soil (Table 1). The vast diversity of microbial species, and their ability to break a wide range of chemical bonds, means that they are responsible for many key soil functions including:

37

L.K. Abbott & D.V. Murphy (eds) Soil Biological Fertility - A Key to Sustainable Land Use in Agriculture. 37-59.

i)   Decomposition of soil organic matter and plant/animal residues with subsequent release of nutrients.

ii) Transformation of compounds between chemical forms; often leading to the formation of more reactive or gaseous compounds which can be lost from the soil.

iii) Degradation of synthetic compounds such as pesticides and herbicides.

iv) Production of antibiotics, which can aid the suppression of soil borne diseases.

v)   Production of soil cementing agents, which may aid aggregation.

vi) Production and degradation of hydrophobic waxy compounds which can lead to water repellence.

vii) Plant nutrient acquisition through symbiotic associations (see Chapter 6 on rhizobia and Chapter 7 on mycorrhizas).

Table 1 Key microbial processes mediating chemical transformations associated with nutrient cycling in soil.

Microbial process	Examples of microbial groups involved
Supply of nutrients	
Mineralisation of organic matter	Heterotrophic microorganisms
Solubilisation of minerals	*Penicillium* sp., *Pseudomonas* sp., *Bacillus* sp.
Nutrient transformations	
Methane (CH_4) oxidation	*Methylococcus* sp., *Methylobacter* sp.
Nitrification	
NH_3 to NO_2^-	*Nitrosospira* sp. and *Nitrosomonas* sp.
NO_2^- to NO_3^-	*Nitrobacter* sp.
Non-symbiotic N_2 fixation	*Azospirillum* sp., *Azotobacter* sp.
Symbiotic N_2 fixation	*Rhizobium* sp., *Anabeana* sp.
Sulphur oxidation	*Thiobacillus* sp., Heterotrophic microorganisms
Loss of nutrients	
CO_2 production	Heterotrophic microorganisms
Methane (CH_4) production	*Methanobacterium* sp., *Methanosarcina* sp.
Denitrification (N_2, N_2O)	*Bacillus* sp., *Pseudomonas* sp., *Agrobacterium* sp.
Reduction of SO_4^{2-} to H_2S	*Desulfovibrio* sp., *Desulfomonas* sp.

Land management practices have considerable impact on the size and dynamics of microbial populations. Intensification of agriculture has focussed on the use of chemical and mechanical inputs, often at the expense of biologically mediated processes. However, even in fertilised systems, microbial processes can play an important role in nutrient supply to plants (Table 2). Where purchased inputs are either costly or unobtainable, microorganisms have a critical role in maintaining soil fertility and crop health (Giller *et al.* 1997). This chapter therefore examines our fundamental understanding of how agricultural management practices and soil amendments influence soil microbial biomass and its activity, and consequently key chemical transformations in soil.

Table 2 Typical rates *(kg ha^{-1} year^{-1})* of soil processes supplying nutrients to crops in temperate agricultural systems and associated typical rates *(kg ha^{-1} year^{-1})* of fertiliser application.

Microbially mediated process	Land use	Soil supply	Reference	Fertiliser[1]
N$_2$ fixation (white clover)	Grassland	13-280	Ladha *et al.* 1992	290
N mineralisation	Grassland	65-400	Jarvis *et al.* 1996	290
S mineralisation	Grassland	18-36	Sakadevan *et al.* 1993	20-32
P mineralisation	Grassland	23	Brookes *et al.* 1984	28
N mineralisation	Arable	50-130	Jarvis *et al.* 1996	200
S mineralisation	Arable	2-6	Kirchmann *et al.* 1996	10-16
P mineralisation	Arable	5	Brookes *et al.* 1984	20

[1] Fertiliser recommendation rates derived from Anon. (2000) where: Grassland = Cut and grazed sward of moderate fertility on medium soil, maintenance application of P; Arable = Dominantly cereal based rotation in moderate rainfall areas on medium soils, maintenance application of P.

2. RELEVANCE OF MICROBIAL DIVERSITY

A single gram of soil contains somewhere in the order of 10^5-10^8 bacteria, 10^6-10^7 actinomycetes and 10^5-10^6 fungal and 10^4 algal colony forming units. Following extraction of soil DNA, Torsvik *et al.* (1994) estimated that one-gram of soil contained several thousand bacterial species. There are probably millions of species of microorganisms within the terrestrial ecosystem but only *ca.* 5% have been identified and/or cultured. With the exception of a few specific populations, our current understanding of microbial functioning has generally been limited to gross estimates of the size and activity of the microbial biomass as a single 'black box' within the soil (see reviews: Jenkinson and Ladd 1981, Wardle 1992, Dalal 1998).

Recent debate in soil research has focused on the importance of microbial diversity in maintaining soil function (Grime 1997, Ritz and Griffiths 2001). Early indications from studies of soil heterotrophic communities across scales of metres to continents showed that despite gross similarities between microbial populations at a coarse scale, at the species to sub-species level, populations showed strong adaptation/evolution to locality (Fulthorpe *et al.* 1998). However, there is currently little direct evidence that links microbial diversity to soil biochemical transformations (Grime 1997, Ritz and Griffiths 2001).

Even where identifiable components of the soil microbial community have been linked to specific transformation processes, e.g. nitrifying bacteria, denitrifying bacteria, methanotrophs, mycorrhizal fungi and *Rhizobium,* there is still limited knowledge of the importance of diversity within these groups on the chemical transformations they mediate. The species composition of microbial populations may be of greater direct relevance to the rate of specific ecosystem processes than their diversity *per se.* Cavigelli and Robertson (2001) found that significantly different N_2O emission rates between soils (differing only in agronomic management) corresponded to differences in denitrifier communities. Less than half of the denitrifying taxa isolated were common to all soils. However, the overall diversity of the denitrifier community in each soil was similar. For processes which are mediated by a range of microbial species, the 'functional diversity' of a microbial population (e.g. the range and complexity of C-substrates that a specific population can decompose) may be of more importance than genetic diversity with respect to ecosystem processes (Zak *et al.* 1994, Tilman *et al.* 1997, Hunter-Cevera 1998). Where resilience to environmental change is critical, the different environmental tolerances, physiological requirements and microhabitat preferences of otherwise functionally similar organisms may also be critical (Perry *et al.* 1989, Yachi and Loreau 1999).

While little is currently known of the factors that favour the development or maintenance of diversity in microbial populations, it seems logical that agricultural management practices which i) conserve or increase the soil microbial biomass, ii) enable niche environments to develop within the soils physical matrix and iii) provide a range of organic compounds on a regular basis will also tend to maintain a diverse microbial population (see review: Kennedy and Gewin 1997).

3.   ENVIRONMENTAL CONTROL

In the soil, microbial communities survive, reproduce and die in a complex 3-dimensional physical framework, which has variable geometry, composition and stability over several orders of magnitude from molecular to field (Foster 1988, Young and Ritz 1998, Ettema and Wardle 2002). Biological activity in soil also changes this framework. The individual effect of environmental and physico-chemical factors on the activity of microorganisms (and hence the breakdown of soil organic matter) has been widely studied (van Veen and Kuikman 1990, Ladd *et al.* 1993, Strong *et al.* 1998, Strong *et al.* 1999). The diversity of the microbial community in soils and the ubiquity of many processes amongst species usually means that chemical transformations have broad optima in relation to environmental conditions. Microbial communities can also adapt to prevailing soil conditions. For example the optimum pH for amino acid breakdown and uptake by communities of soil bacteria was shown to be very close to the natural pH of the soils from which they were extracted (Bååth 1996).

Soil texture is a primary mediating factor in the operation of external climatic conditions in the soil through its influence on the retention of soil organic matter and development of soil structure and pore size distribution. Good relationships are often found between clay content and microbial biomass (Figure 1), although this is also a result of the relationship between clay content and soil organic matter (Wardle 1992).

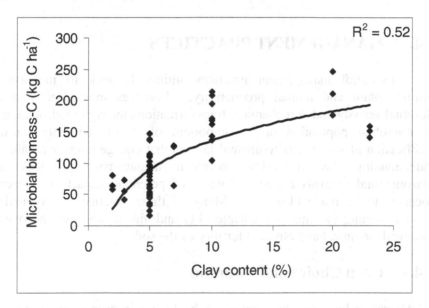

Figure 1 Relationship between clay content and microbial biomass in coarse-texture agricultural soils. Soils (0-5 cm) were collected under winter wheat from a single catchment (10 x 20 km; 450 mm annual precipitation; 0.4 to 2.7 %C). Differences in microbial biomass for a specific clay content reflect the associated range of soil organic C contents resulting from prior differences in crop rotations and farm management practices. D.V. Murphy and N. Milton (unpublished data).

Well-structured finer textured soils also create more niche environments for microbial colonisation. This generally supports greater microbial populations due to protection from desiccation (Bushby and Marshall 1977) and predators (Roper and Marshall 1978) and because of the formation of different oxygen concentrations. For example, Arah and Smith (1989) showed how microbial respiration and soil water interact to create a complex and spatially heterogeneous network of aerobic and anaerobic microsites, resulting in simultaneous nitrification and denitrification.

Microbial activity is often strongly positively related to changes in temperature under field conditions throughout the cropping season when moisture is not limited (Campbell *et al.* 1999). However, external environmental conditions do not apply uniformly throughout the soil and because of the complex dynamics of soil ecosystems, no single parameter is satisfactory as an indicator of microbial activity in soils under different conditions (Dick 1992). Consequently Wardle (1992) showed vastly different responses of the microbial biomass to seasonal climatic variability, even in very similar ecosystems, resulting from the complexity of interactions between soil moisture, temperature and their effects on plant growth.

4. MANAGEMENT PRACTICES

Agricultural management practices ultimately seek to increase or optimise plant and animal productivity. Practices may directly affect microbial activity and soil chemical transformations through modification of the microbial population or a component of it, or indirectly through modification of soil or environmental factors that change microbial habitats. Understanding the interactions between management practices and environmental controls allows estimates of potential impacts on specific processes to be made (Table 3). Many of these practices are typical of modern farming systems (see Chapter 11) and aim to improve the overall biological, chemical and physical fertility of the soil.

4.1 Crop Choice and Rotation

Microbial biomass also appears to be higher in crop rotations than in monocultures (e.g. Anderson and Domsch 1989, Moore *et al.* 2000), which probably reflects the greater niche diversity provided by a wider range of inputs to the soil in crop residues. Not all rotations increase soil organic matter levels and microbial biomass to the same extent. Edwards *et al.* (1990) showed that addition of soybean to a crop rotation did not increase either microbial biomass or activity. However, adapting arable rotations to include green manure, cover crops or short-term pastures will increase the duration of crop cover, and the amount and diversity of crop residues. For example, Campbell *et al.* (1991) showed that changing from a monoculture of spring wheat to a rotation containing a legume or green manure increased the proportion of the year under plant cover, increased returns of C to the soil and resulted in an increase in microbial biomass. Murphy *et al.* (1998) showed that increasing the proportion of legume in the rotation increased total soil N, microbial biomass N and the seasonal cumulative gross N mineralisation rate (Table 4).

The choice and order of crops within a rotation is made under a number of constraints (e.g. market availability, weed and disease control). It is clear that management of crop choice and rotation has both direct and indirect impacts on the activity and diversity of soil microorganisms (Figure 2). However, how these choices can be practically manipulated to benefit the microbial biomass (or particular components) and microbially mediated chemical transformations, except at the coarsest level, is not yet understood.

Table 3 Likely direct impact of agricultural management practices on the microbial biomass and key C and N chemical transformation processes in an arable cropping system. ↑ = increase in pool/process, ↓ = decrease in pool/process.

	Microbial biomass	Mineralisation	Nitrification	CO$_2$ emissions	Methane production	Denitrification
Crop rotation including grassland or green manures instead of continuous arable cropping.	↑	↑		↑		
Retention of crop residues instead of burning.	↑	↑		↑		
Minimum tillage practices instead of full soil cultivation techniques.	↑	↓		↓	↓	↑
Irrigation of crops in rain-limited cropping environments.	↑	↑	↑	↑		↑
Drainage of agricultural land in high rainfall/waterlogged environments.		↑	↑	↑	↓	↓
Application of fungicides to soil.	↓					
Application of inorganic N fertilisers to soil.			↑			↑
Application of organic amendments to soil.	↑	↑		↑		↑
Liming of soil to raise pH on acidic soils.	↑	↑	↑	↑		↑

Table 4 Pools (kg N ha^{-1}) and fluxes (kg N ha^{-1} cycled during wheat crop) of N in contrasting cropping systems (0-5 cm) in Western Australia (adapted from Murphy *et al.* 1998).

	Continuous wheat	Lupin -wheat rotation	Continuous subterranean clover
Total organic N pool	1008	1002	1463
Microbial biomass N pool	64	68	76
Gross N mineralisation flux	100	120	282
Gross N immobilisation flux	57	61	160
Net N mineralisation flux	43	59	122
N flux through microbial biomass	25	41	49

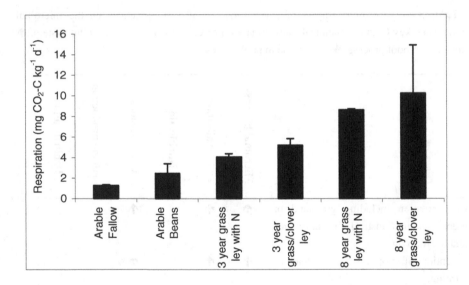

Figure 2 Influence of crop rotation on soil microbial activity (0-23 cm). Data collected in 1996 from the Rothamsted Continuous Ley-Arable experiment UK (est. 1938). Capped bars are standard errors of the mean. D.V. Murphy and P.R. Poulton (unpublished data).

4.2   Crop Residue Management

Microbial activity is generally C-limited in agricultural soils. Retention of crop residues therefore provides a practical means of increasing soil microbial populations without having to import additional organic matter. Management of crop residues changes soil temperature, water and the distribution of plant residues and organic matter which all influence the location and activity of microorganisms (e.g. Hendrix *et al.* 1986; see

Chapter 10). Surface mulching of crop residues increased populations of bacteria, actinomycetes and fungi 2-6 times compared to non-mulched treatments (Doran 1980). In contrast, stubble burning reduces inputs of organic matter and increases soil temperature and moisture deficits after harvest and tends to reduce the population and activity of soil microorganisms (e.g. Thompson 1992). For example, Powlson *et al.* (1987) showed that incorporation of straw instead of burning, increase the microbial biomass by 45%, although total C was only 5% higher.

The decomposition of crop residues in soils is largely controlled by their chemical composition, residue management practices (e.g. incorporation method), soil and environmental factors (see review: Kumar and Goh 2000). Residue quality, which controls decomposability and nutrient release (N, P and S), is mainly determined by residue composition i.e. contents of soluble carbohydrates, amino acids, active polyphenols, lignin, nutrients and C:nutrient ratio (Handayanto *et al.* 1995). The decomposition of crop residues in soils can be partly regulated through the control of the quality of litter inputs. Both crop selection and the mixing of crop residues from separate sources before incorporation have been used to manage nutrient release during microbial decomposition (Palm *et al.* 1997). It is well known that lignin in plant residues slows microbial decomposition and it is possible to breed plants with higher levels of lignin (Paustian *et al.* 1995). Developments in plant genetic engineering have resulted in the modification of lignin structure to improve wood quality (in pulp and paper production) and for crop digestibility (Baucher *et al.* 1998). Manipulation of plant residue quality in this way has significant future implications for controlling soil organic matter formation and subsequent rates of mineralisation.

4.3　Tillage

The greater the intensity of energy input to the soil through tillage the greater the rates of residue decomposition (Watts *et al.* 2000) which regulates the size of the microbial population (Figure 3). Mixing of crop residues and soil also favours the development of bacterial rather than fungal populations and alters both the pathways of decomposition and the network of predators that develop in soil (McGonigle and Miller 1996). Martens (2001) reviewed 21 papers reporting the effects of tillage systems on soil microorganisms and their activity. All tillage systems dramatically reduced the microbial biomass relative to uncultivated native vegetation; the studies reported had high levels of soil organic matter and microbial biomass. However, as tillage intensity decreased, declines in microbial biomass were less (Martens 2001).

In reduced tillage and direct drill systems, microorganisms and their associated processes become concentrated in the surface soil and the proportion of organic carbon immobilised in the microbial biomass is higher than in cultivated soils with similar microbial biomass levels (Wardle 1992). Minimum cultivation practices result in less surface area contact between residues and soil microbial communities reducing decomposition rates; such conditions also favour the development of stable fungal populations during residue decomposition (Martens 2001). In dryland cropping systems, where the soil surface dries out rapidly, colonisation of surface residue by the microbial population may be restricted by water. In such cases, the mulching effect of the residue (i.e. surface area) may be of greater importance to the rate of residue decomposition than its quality/composition (Sparling *et al.* 1995).

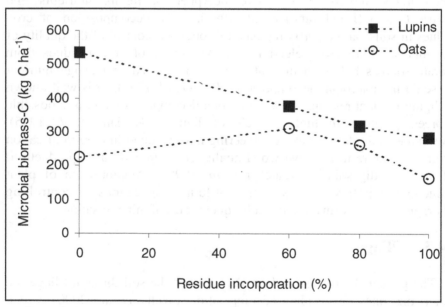

Figure 3 Effect of degree of residue (lupins or oats) incorporation by tillage on the level of microbial biomass in the 0 - 30 cm soil layer of a loamy sand: 0% = brown manure; 60% = offset disc to 10 cm; 80% = disc plough to 15 cm and 100% = molboard plough to 25 cm. Crops were incorporated as a green manure except for the 0% residue incorporation treatment where the crop was killed with chemicals. D.V. Murphy, F. Hoyle and N. Milton (unpublished data).

Minimum tillage and direct drill systems also lead to changes in many of the environmental factors that control microbial activity and residue breakdown. Higher moisture levels, lower O_2 concentrations and reduced soil temperatures result in lower early season microbial activity and lower mineralisation of N in temperate climates despite the higher microbial biomass populations in no till systems (Martens 2001). The altered physical conditions also favour higher losses of N through denitrificatio. Aulakh *et al.* (1984) measured gaseous losses of 12-16 kg N ha^{-1} year^{-1} in no till compared to 3-7 kg N ha^{-1} year^{-1} in conventionally tilled soils.

It is clear that reducing tillage intensity can increase the size of the microbial biomass population, where crop residues are retained in the system, and lead to a dominance of fungi over bacteria within the microbial biomass (at least in the surface horizons of the soil). Reduced tillage systems therefore have significantly different food webs than conventional intensively tilled systems (Verhoef and Brussaard 1990). However, it is not clear whether changes in community structure or modification in the environmental factors controlling microbial processes has the larger effect on microbially mediated chemical transformations. Where possible, optimisation of tillage systems for soils and rotations should seek to minimise soil disturbance, reduce water and wind erosion, maintain soil organic matter levels and protect soil microbial/faunal populations. However, it is the interactions between tillage and residue management that are critical in the management of microbially mediated chemical transformations in soils.

4.4 Irrigation and Drainage

The main effect of irrigation and drainage on the soil microbial biomass is indirect, by regulating the seasonal effects of rainfall and tending to stabilise the soil moisture regime throughout the growing season. Other indirect effects occur as the duration of crop growth and consequent inputs of organic matter through exudates and crop residues tend to increase when irrigation is used. Although these effects have been shown to be important, they are generally smaller than those caused by climate and residue quality (Wilkinson *et al.* 2002). Irrigation (where practical) is used when crop yields or quality would otherwise be compromised by drought. Soil conditions will therefore be maintained at optimum moisture contents for microbial activity for longer periods of the cropping season. However, this may reduce the turnover of nutrients through the microbial biomass during the growing season, because fewer wet-dry cycles will occur (Wardle 1992). Poorly managed irrigation systems may also lead to increases in soil salinity. This has been shown to reduce microbial enzyme activity and respiration (Garcia and Hernandez 1996). Soil drainage increases soil aeration, reducing losses of N by denitrification, and tends to increase the seasonal duration of microbial activity including nitrification (Table 5). Coupled with increased throughflow of water, this may increase leaching losses of N, as nitrate and hence merely trade one form of N loss for another.

Table 5 Comparison of C and N transformation rates in adjoining undrained and drained grassland soils. * Denotes significant (P<0.10) difference between undrained and drained soils for a given parameter. MB-N = microbial biomass-N. D.V. Murphy (unpublished data).

	Total C	MB-N	CO_2-C	Gross N fluxes Mineralisation	Nitrification
	%	mg kg^{-1}		mg kg^{-1} d^{-1}	
Undrained	7.39*	338	15.9	1.66	0.01
Drained	6.29	328	18.4*	1.81*	0.32*

4.5 Pesticides

Microorganisms play an important role in the biodegradation of pesticides, herbicides and other environmental contaminants. Biodegradation may occur through a series of independent reactions; the direct interaction of a number of microorganisms in a consortium is then required for complete degradation to occur. Enhanced degradation rates (i.e. reduced microbial lag phase) of chemicals can occur after repeated applications to the same soil (Roeth 1986, Smith and Lanfond 1990) as a result of preferential selection of the microorganisms/biochemical pathways involved. This adaptation increases the rate of biodegradation and is so effective in some instance that the efficacy of the pesticide (Roeth 1986, Felsot 1989) or herbicide (Audus 1949) is reduced sufficiently to limit plant productivity.

The direct effect of pesticides on soil organisms depends on the type and specificity of the chemical, the susceptibility of target species and the rate of application (van der Werff 1996). Fungicides usually exert predictable negative effects on the fungal component of the soil microbial biomass; nematicides and soil fumigants may also cause a significant temporary reduction in the soil microbial biomass. Herbicides and insecticides tend to have smaller and more variable effects. Hart and Brookes (1996) showed that the continuous use of five pesticides (either singly or in combination) had no measurable long-term harmful effects on the soil microbial biomass or its activity. The addition of aldicarb (2-methyl-2-(methylthio)propionaldehyde *O*-methylcarbamoyloxime) even increased microbial biomass carbon by 7-16%. The increasing specificity and reduced doses of active ingredients in new pesticide formulations has resulted in negative effects of pesticides on soil microorganisms to be recorded more rarely.

4.6   Inorganic Fertilisers

Mineral fertilisers (usually simply inorganic salts) are primarily used to overcome the nutrient limitations to plant growth, which occur in many farming systems. They may also be used to optimise product quality. The response of the microbial biomass to fertiliser amendments depends critically on whether the nutrient applied is limiting microbial activity or growth. Additionally at high rates of fertiliser addition, osmotic effects may occur, especially in zones close to fertiliser granule or liquid injection points. However, such effects are only temporary and often only at rates of fertiliser addition well in excess of normal farm rates. High concentrations of soluble nutrients in the soil have been shown to discourage the activity of some symbiotic microorganisms. For example, N_2 fixation by rhizobia is significantly reduced in soils with high concentrations of mineral N. Similarly high concentrations of soluble P in soil solution suppress the infectivity of mycorrhizal fungi to their host plants.

Mineral N applications have been shown, depending on the study, to either increase or decrease microbial biomass. However, around half of all studies published show only very slight effects (Wardle 1992). Application of anhydrous ammonia initially kills many soil microorganisms; bacteria and actinomycete populations recover within 1 to 2 weeks. However the fungal population may take as long as 7 weeks to recover (Doran and Werner 1990). Long-term applications of ammonium fertiliser to grassland caused a significant decrease in methane oxidation rates, but the application of nitrate for the same length of time did not (Willison *et al.* 1995, Table 6). In addition, a negative correlation between short-term methane uptake and the rate of ammonium applied was measured for arable and grassland soils in the laboratory (Tlustos *et al.* 1998). The use of ammonium fertiliser may also favour the development of a population of soil nitrifiers, which outcompete methanotrophs for niche environments within the soil. While nitrifiers are also able to oxidise methane, their rates of oxidation are significantly slower than those of methanotrophs (Bedard and Knowles 1989).

Targeted chemicals can be used as a fertiliser coating or applied directly to the soil to inhibit nitrification (usually by competition for the active site of the ammonium mono-oxygenase enzyme; McCarty 1999). By maintaining added N in the ammonium form for an extended period, losses can be reduced, plant uptake efficiency increased and microbial immobilisation of applied N may be increased. Use of nitrification inhibitors does not seem to have any long-term negative effect on the population of nitrifying bacteria in soil (apart from reducing N availability). There has been little uptake of such technologies in conventional agricultural practice, however, as the agronomic benefits often do not outweigh their cost. Environmental concerns of nitrate leaching may lead to the use of nitrification inhibitors in specific regions in the future.

Table 6 Pools and rates of microbially mediated nutrient transformations in plots of the Broadbalk continuous wheat experiment, Rothamsted Research UK. The soil is a calcareous silty clay loam. Plots 8 and 16 receive maintenance applications of P, K and Mg.

	No N added	144 kg N ha^{-1} y^{-1}	288 kg N ha^{-1} y^{-1}	Farmyard manure
	Plot 3	Plot 8	Plot 16	Plot 2.2
[a] Wheat yield (t ha^{-1} at 85% dry matter)	0.93	6.01	8.22	4.93
[a] Estimated return in stubble and chaff (t ha^{-1})	0.6	2.3	3.1	1.9
[a] pH (H$_2$O)	8.2	7.3	7.8	7.8
[a] Soil organic C (%)	0.76	1.06	1.16	3.00
[a] Soil organic N (%)	0.09	0.12	0.13	0.31
[a] Olsen P (mg kg^{-1})	7	78	75	102
[b] Microbial C (mg kg^{-1})	167	259	nd	545
[c] Microbial P (mg kg^{-1})	6.0	5.3	nd	28.9
[d] CO$_2$ production (mg CO$_2$-C kg^{-1} d^{-1})	7.2	8.5	9.5	13.8
[d] Gross N mineralisation (mg N kg^{-1} d^{-1})	0.08	0.23	0.39	0.89
[e] Gross nitrification (mg N kg^{-1} d^{-1})	0.16	0.21	nd	0.76
[d] Methane oxidation (μg CH$_4$-C kg^{-1} d^{-1})	0.50	0.43	0.24	0.53
[e] Nitrifier populations gene copies g^{-1}	1.3 x 10^4	5 x 10^5	nd	1.2 x 10^5

[a] P.R. Poulton (unpublished data). Data from 1997 except wheat yield which is the average of 1996-2000. [b] Wu (1991) PhD thesis. [c] Brookes *et al.* (1984). [d] D.V. Murphy (unpublished data). [e] Mendum *et al.* (1999). nd = not determined.

Fertilisers can also have indirect effects on soil microorganisms through effects on plant growth. For example, optimisation of potassium (K) fertilisation can have a stimulatory effect on microbial activity through increases in root exudation, which supply energy to the microbial biomass. The use of fertilisers increase plant yield and thus increases the return of C to soil in above- and below-ground plant residues, which increase soil microbial activity (see review; Dick 1992). On the Broadbalk Continuous Wheat Experiment (started in 1843) long-term inorganic N applications resulted in a small increase in soil organic matter, heterotrophic microbial activity and microbial biomass compared to plots that have never received N fertiliser

(Table 6). However, rates of gross N mineralisation are significantly higher where larger rates of N fertiliser have been applied (Table 6).

4.7 Organic Fertilisers

Many cropping systems use the waste products of livestock enterprises as part of their fertilisation strategy either as solid materials (manure) or in liquid form (slurry). Other organic materials e.g. sewage sludge, food processing wastes and composts of various materials are also applied to soils depending on local availability. The application of organic materials tends to stimulate the microbial biomass directly and substantially, unlike applications of mineral fertilisers (Wardle 1992, Table 6). Organic fertilisers add C, N and other nutrients simultaneously satisfying components of the microbial population which are (usually) otherwise C-limited. Continuing release of nutrients as the materials are slowly decomposed in soils can also sustain the microbial biomass population for longer periods of time compared to the impact of mineral fertilisers e.g. the maintenance of a higher population of nitrifiers where farmyard manure rather than mineral fertilisers were applied in the Broadbalk experiment (Table 6).

Achieving synchrony between crop demand and nutrient supply is very difficult (Myers *et al.* 1997) particularly where nutrients are supplied solely through microbially mediated chemical transformations and decomposition of a diverse and variable range of organic materials. However, optimum efficiency of nutrient use seems to be achieved in many situations where mineral fertilisers and organic materials are used together in fertilisation strategies for cropping systems (Palm *et al.* 1997). There is some evidence that the mineral N pool applied in manures is more efficiently used for plant uptake than mineral N fertiliser (Stockdale *et al.* 1995). Also a large proportionate increase in both microbial P and the conventionally measured forms of available P is measured when farmyard manure is applied over the long-term (Table 6). In soils, which strongly fix inorganic P, combined use of soluble phosphate fertilisers with manure stimulates the uptake of P by the biomass, thus protecting it from immediate fixation and significantly increasing crop yield (Twomlow *et al.* 1999). Incorporation of the mineral N and P added in manure into the microbial biomass through immobilisation may protect these nutrients from loss before crop roots are fully developed and release of nutrients through microbial biomass turnover and predation may be more closely matched to crop demand.

Application of manures and/or sewage sludge to soils can significantly increase the heavy metal loading of soils. For example, Brookes and McGrath (1984) demonstrated that heavy metals derived from sewage sludge substantially reduced microbial biomass even twenty years after application.

The reduced microbial activity where organic materials are repeatedly applied leads to extreme accumulation of organic matter in the soil (Chander and Brookes 1991). Also, heavy metals decrease microbial diversity as a result of species extinction due to a lack of tolerance to the imposed stress and/or the competitive advantage of certain species who predominate in the presence of the heavy metal stress (Giller *et al.* 1998). Heavy metal content of organic materials therefore has significant implications for the use and management of such materials in agricultural systems.

4.8   Other Soil Amendments

Liming of soils to counteract acidity caused by atmospheric deposition, fertiliser addition or natural mineral weathering is widely practiced to optimise soil pH and other conditions for crop growth. Consequently, liming also increases both microbial biomass and specific populations (e.g. nitrifiers). Liming also creates soil conditions that favour the microbial processes that degrade hydrophobic substances around sand grains and thus reduce the water repellency of sandy soils (Roper 1998) and increase the populations of wax-degrading bacteria (mainly actinomycetes).

Amending clay soils with sand and sandy soils with clay (particularly calcareous clays, i.e. marling) has been practiced for many centuries worldwide to improve crop nutrition and soil structure (e.g. Piggott 1981). Clay amendment still occurs today (e.g. on infertile sandy soils in Australia) and has been shown to increase crop yields and thus plant above- and below-ground returns, which affect the microbial biomass indirectly, as discussed above. Clay amendment also ameliorates the non-wetting characteristics of sandy soils, which is likely to lead to improved water holding capacity and protects microorganisms from desiccation (Bushby and Marshall 1977) thus improving conditions for microbial activity (Marshall 1975).

Applications of glucose and molasses to soil are currently being promoted as a means of 'feeding the microbes' directly and enhancing general soil biological fertility as the additional C is cycled through the soil food web. Daly and Stewart (1999) found that CO_2 evolution increased after addition of molasses to soil but there was no corresponding increase in mineralisation of N, S or P, i.e. the molasses was rapidly broken down without any additional stimulation of microbial decomposition of organic matter in the soil. C addition rates are often low (10-50 kg C ha^{-1}) and given that 40-60% of the organic C is respired during microbial decomposition, little C is directly assimilated by the microorganisms (Wu 1991). As a result, regular applications are likely to be required to enhance microbial activity and cause significant microbial immobilisation of nutrients. Humic acids can also be added to soils where they are degraded and/or transformed by soil microorganisms as a supplemental source of organic carbon (see review; Filip *et al.* 1998).

4.9 Microbial Inoculants

Inoculants have been used successfully to enhance a limited number of microbial populations in soil. Most notably *Rhizobium* inoculation to legume crops to maximise plant N uptake through symbiotic biological N_2 fixation has proved to be highly successful (see Chapter 6). By 1958, 10^7 hectares in the former Soviet Union had been treated with various bacterial inocula, increases in yield of 10-20% were reported in 50-70% of trials and similar results have also been obtained elsewhere (Kloepper *et al.* 1989). Phosphate-solubilising microorganisms (e.g. *Aspergillus, Penicillium*) have long been promoted to increase P uptake by plants (see review; Whitelaw 2000). Free-living N_2 fixing microorganisms (e.g. *Azotobacter*) have also been applied to the soil to increase plant N availability. Soil amendments, with more generalised microbial populations, are still widely used (Daly and Stewart 1999, Kinsey and Walters 1999). However, it is often difficult to determine the causal link between the introduction of beneficial microorganisms and product claims. Increases in plant yield associated with such inoculants have also been partly attributed to the enhancement of root growth by bacterial phytohormone production and/or nutrients supplied by the turnover of the added inoculum itself (Bashan and Holguin 1997, de Freitas *et al.* 1997).

The performance of microbial inocula in trials is very inconsistent. This is not surprising given the complexity of the relationships between inoculants, indigenous populations, crops, climate and the soil matrix. The major constraint to the successful introduction of microbial inoculants to soil is the ability of an introduced microbial population to exist at levels above indigenous population numbers (Hirsch 1996). In most cases it will be unlikely that introduced microorganisms will be more suited than the indigenous population, which has already co-evolved to cope with localised environmental conditions (e.g. substrate, temperature, prolonged drying, periodic waterlogging, pH) and management pressures (land use, tillage *etc.*). Greater success is likely with the introduction of endophytes, which have less competition within soil roots. A fundamental understanding of the growth and survival characteristics of the microorganisms along with knowledge of how they are likely to interact with the indigenous microbial population and plant species (i.e. crop rotation) will aid in successful introduction of specific microbial inoculants.

5.   CONCLUSION

Many of the chemical transformations, which occur in soil, are mediated by the soil microbial biomass. Microorganisms therefore have a major controlling influence on the cycling and loss of nutrients in soil and the regulation of plant nutrient availability, as well as many other transformation processes. The quantity and quality of organic matter in soil is a major factor in controlling the abundance and activity of microorganisms and therefore underpins many of the microbially mediated chemical transformations in soil.   While the direct and indirect effects of common agricultural management practices on the soil microbial biomass are known, it is less clear which combination of locally adapted management practices are able to maintain and increase soil organic matter status and thus likely to optimise the activity of the microbial biomass in soil (see Chapter 11). At the coarsest level it can be readily advised that minimum appropriate soil tillage with regular incorporation of a diverse range of crop residues and other organic materials is the key to supporting enhanced microbial activity and associated chemical transformations in agricultural soils.

6.   ACKNOWLEDGEMENTS

DVM receives funding through the Australian Grains Research and Development Corporation and The University of Western Australia. Rothamsted Research receives grant-in-aid form the UK Biotechnology and Biological Sciences Research Council.

7.   REFERENCES

Anderson T H and Domsch K H 1989 Ratios of microbial biomass carbon to total organic carbon in arable soils. Soil Biology and Biochemistry 21: 471-480.
Anon. 2000 Fertiliser Recommendations for Agricultural and Horticultural Crops. UK Ministry of Agriculture Fisheries and Food Reference Book 209. pp. 177. Stationery Office. London.
Arah J R M and Smith K A 1989 Steady-state denitrification in aggregated soils - a mathematical model. Journal of Soil Science 40: 139-149.
Audus L J 1949 Biological detoxification of 2,4-D. Plant and Soil 3: 170-192.
Aulakh M S Rennie D A and Paul E A 1984 Gaseous nitrogen losses from soils under zero till as compared with conventional-till management systems.   Journal of Environmental Quality 13: 130-136.
Bååth E 1996 Adaptation of soil microbial communities to prevailing pH in different soils. FEMS Microbiolgy Ecology 19: 227-237.

Bashan Y and Holguin G 1997 *Azospirillim*-plant relationships: Environmental and physiological advances. Canadian Journal of Microbiology 43: 103-121.

Baucher M, Monties B, Van Montagu M and Boerjan W 1998 Biosynthesis and genetic engineering of lignin. Critical Reviews in Plant Sciences 17: 125-197.

Bedard C and Knowles R 1989 Physiology, biochemistry and specific inhibitors of CH_4, NH_4^+ and CO oxidation by methanotrophs and nitrifiers. Microbiological Reviews 53: 68-84.

Brookes P C and McGrath S P 1984 Effects of metal toxicity on the size of the soil microbial biomass. Journal of Soil Science 35: 341-346.

Brookes P C, Powlson D S and Jenkinson D S 1984 Phosphorus in the soil microbial biomass. Soil Biology and Biochemistry 16: 169-175.

Bushby H V A and Marshall K C 1977 Some factors affecting the survival of root-nodule bacteria on desiccation. Soil Biology and Biochemistry 9: 143-147

Campbell C A, Biederbeck V O, Zentner R P and Lafond G P 1991 Effect of crop rotation and cultural practices on soil organic matter, microbial biomass and respiration in a thin black Chernozem. Canadian Journal of Soil Science 71: 363-376.

Campbell C A, Lafond G P, Biederbeck V O, Wen G, Schoenau J and Hahn D 1999 Seasonal trends in soil biochemical attributes: effects of crop management on a black chernozem. Canadian Journal of Soil Science 79: 85-97.

Cavigelli M A and Robertson G P 2001 Role of denitrifier diversity in rates of nitrous oxide consumption in a terrestrial ecosystem. Soil Biology and Biochemistry 33: 297-310.

Chander K and Brookes P C 1991 Effects of heavy metals form past applications of sewage sludge on microbial biomass and organic matter accumulation in a sandy loam and a silty loam UK soil. Soil Biology and Biochemistry 23: 927-932.

Curl E A and Truelove B 1986 The rhizosphere. Springer-Verlag. Berlin.

Dalal R C 1998 Soil microbial biomass – what do the numbers really mean? Australian Journal of Experimental Agriculture 38: 649-665.

Daly M J and Stewart D P C 1999 Influence of 'effective microorganisms' (EM) on vegetable production and carbon mineralisation – A preliminary investigation. Journal of Sustainable Agriculture 14: 15-25.

de Freitas J R, Banerjee M R and Germida J J 1997 Phosphate-solubilizing rhizobacteria enhance the growth and yield but not phosphorus uptake of canola (*Brassica napus* L.). Biology and Fertility of Soils 24: 358-364.

Dick R P 1992 A review - long-term effects of agricultural systems on soil biochemical and microbial parameters. Agricultural Ecosystems and Environment 40: 25-36.

Doran J W and Werner M R 1990 Management and soil biology. *In*: Sustainable Agriculture in Temperate Zones. C A Francis, C B Flora, L D King (eds.) pp. 151-177. Wiley. New York.

Doran J W 1980 Microbial changes associated with residue management with reduced tillage. Soil Science Society of America Journal 44: 518-524.

Edwards C A, Lal R, Madden P, Miller R H and House G 1990 Sustainable Agricultural Systems. Soil and Water Conservation Society. Ankeny, Iowa.

Ettema C H and Wardle D A 2002 Spatial soil ecology. Trends in Ecology and Evolution 17: 177-183.

Felsot A S 1989 Enhanced biodegradation of insecticides in soil: implications for agroecosystems. Annual Reviews in Entomology 34: 453-476.

Filip Z, Claus H and Dippell G 1998 Degradation of humic substances by soil microorganisms – a review. Zeitschrift fur Pflanzenernahrung und Bodenkunde 161: 605-612.

Foster R C 1988 Microenvironments of soil microorganisms. Biology and Fertility of Soils 6: 189-203.

Foster R C, Rovira A D and Cook T W 1983 Ultrastructure of the root-soil interface. The American Phytopathology Society. St. Paul, Minnesota, USA.

Fulthorpe R R, Rhodes A N and Tiedje J M 1998 High levels of endemicity of 3-chlorobenzoate-degrading soil bacteria. Applied Environmental Microbiology 64: 1620-1627.

Garcia C and Hernandez T 1996 Influence of salinity on the biological and biochemical activity of a calciorthid soil. Plant and Soil 178: 255-263.

Giller K E, Beare M H, Lavelle P, Izac A-M N and Swift M J 1997 Agricultural intensification, soil biodiversity and agroecosystem function. Applied Soil Ecology 6: 3-16.

Giller K E, Witter E and McGrath S P 1998 Toxicity of heavy metals to microorganisms and microbial processes in agricultural soils: a review. Soil Biology and Biochemistry 30: 1389-1414.

Grayston S J, Vaughan D and Jones D 1996 Rhizosphere carbon flow in trees, in comparison with annual plants: the importance of root exudation and its impact on microbial activity and nutrient availability. Applied Soil Ecology 5: 29-56.

Grime J P 1997 Biodiversity and ecosystem function: the debate deepens. Science 277: 1260-1261.

Handayanto E, Cadisch G and Giller K E 1995 Manipulation of quality and mineralisation of tropical legume tree prunings by varying nitrogen supply. Plant and Soil 176: 149-160.

Hart M R and Brookes P C 1996 Soil microbial biomass and mineralisation of soil organic matter after 19 years of cumulative field applications of pesticides. Soil Biology and Biochemistry 28: 1641-1649.

Hendrix P E, Parmalee R W, Crossby Jr D A, Coleman D C, Odum E P and Groffman P M 1986 Detritus food webs in conventional and no-tillage agroecosystems. BioScience 36: 374-380.

Hirsch P R 1996 Population dynamics of indigeneous and genetically modified rhizobia in the field. New Phytologist 133: 159-171.

Hunter-Cevera J C 1998 The value of microbial diversity. Current Opinion in Microbiology 1: 278-285.

Jarvis S C, Stockdale E A, Shepherd M A and Powlson D S 1996 Nitrogen mineralisation in temperate agricultural soils: Processes and measurement. Advances in Agronomy 57: 187-234.

Jenkinson D S and Ladd J N 1981 Microbial biomass in soil: Measurement and turnover. In: Soil Biochemistry Volume 5. E A Paul and J N Ladd (eds.) pp. 415-471. Marcel Dekker. New York.

Kennedy A C and Gewin V L 1997 Soil microbial diversity: Present and future considerations. Soil Science 162: 607-617.

Kinsey N and Walters C 1999 Neal Kinsey's Hands-on Agronomy. Acres USA.

Kirchmann H, Pichlmayer F and Gerzabek M H 1996 Sulfur balances and sulfur-34 abundance in a long term fertiliser experiment. Soil Science Society of America Journal 60: 174-178.

Kloepper J W, Lifshitz R and Zablotowicz R M 1989 Free-living bacterial inocula for enhancing crop productivity. Trends in Biotechnology 7: 39-44.

Kumar K and Goh K M 2000 Crop residues and management practices: Effects on soil quality, soil nitrogen dynamics, crop yield, and nitrogen recovery. Advances in Agronomy 68: 197-319.

Ladd J N, Foster R C and Skjemstad J O 1993 Soil structure: carbon and nitrogen metabolism. Geoderma 56: 401-434.

Ladha J K, George T and Bohlool B B (eds.) 1992 Role of biological nitrogen fixation in sustainable agriculture. Plant and Soil 141: 1-209.

Marshall K C 1975 Clay mineralogy in relation to survival of soil bacteria. Annual Review of Phytopathology 13: 357-373.

Martens D A 2001 Nitrogen cycling under different soil management systems. Advances in Agronomy 70: 143-192.

McCarty G W 1999 Modes of action of nitrification inhibitors. Biology and Fertility of Soils 29: 1-9.

McGonigle T P and Miller M H 1996 Development of fungi below ground in association with plants growing in disturbed and undisturbed soils. Soil Biology and Biochemistry 3: 263-269.

Mendum T A, Sockett R E and Hirsch P R 1999 Use of molecular and isotopic techniques to monitor the response of autotrophic ammonia-oxidising populations of the β subdivision of the class *Proteobacteria* in arable soils to nitrogen fertiliser. Applied and Environmental Microbiology 65: 4155-4162.

Moore J M, Klose S and Tabatabai M A 2000 Soil microbial biomass carbon and nitrogen as affected by cropping systems. Biology and Fertility of Soils 31: 200-210.

Murphy D V, Fillery I R P and Sparling GP 1998 Seasonal fluctuations in gross N mineralisation, ammonium consumption and microbial biomass in a Western Australian soil under different land uses. Australian Journal of Soil Research 49: 523-535.

Myers R J K, van Noordwijk M and Vityakon P 1997 Synchrony of nutrient release and plant demand: Plant litter quality, soil environment and farmer management options. *In:* Driven by Nature: Plant litter quality and decomposition. G Cadisch and K E Giller (eds.) pp. 215-229. CAB International. Wallingford, Oxon, UK.

Neal J L Jr, Larson R I and Atkinson T G 1973 Changes in rhizosphere populations of selected physiological groups of bacteria related to substitution of specific pairs of chromosomes in spring wheat. Plant and Soil 39: 209-212.

Palm C A, Myers R J K and Nandwa S 1997 Organic-inorganic nutrient interactions in soil fertility replenishment. *In:* Replenishing Soil Fertility in Africa. R J Buresh, P A Sanchez and F G Calhoun (eds.) ASA-SSSA Special Publication 51: 193-218.

Paustian K, Robertson G P and Elliot E T 1995 Management impacts on carbon storage and gas fluxes (CO_2, CH_4) in mid-latitude cropland. *In:* Soil Management and Greenhouse Effect. R Lal, J Kimble, E Levine and B A Stewart (eds.) pp. 69-83. Advances in Soil Science. Lewis Publishers. Boca Raton, USA.

Perry D A, Amaranthus M P, Borchers J G, Borchers S L and Brainerd R E 1989 Bootstrapping in ecosystems. Bioscience 39: 230-237.

Piggott S 1981 The Agrarian History of England and Wales Volume 1. I. Prehistory. pp. 213-214. Cambridge University Press. Cambridge.

Powlson D S, Brookes P C and Christensen B T 1987 Measurement of microbial biomass provides an early indication of changes in total soil organic matter due to straw incorporation. Soil Biology and Biochemistry 19: 159-164.

Ritz K and Griffiths B S 2001 Implications of soil biodiversity for sustainable organic matter management. *In:* Sustainable management of soil organic matter. R M Rees,

B C Ball, C D Campbell and C A Watson (eds.) pp. 343-356. CAB International. Wallingford, Oxon, UK.

Roeth F W 1986 Enhanced herbicide degradation in soil with repeat application. Reviews in Weed Science 2: 45-65.

Roper M 1998 Sorting out sandy soils. Microbiology Australia 19: 6-7.

Roper M M and Marshall K C 1978 Effects of a clay mineral on microbial predation and parasitism of *Escherichia coli*. Microbial Ecology 4: 279-289.

Sakadevan K, MacKay A D and Hedley M J 1993 Sulphur cycling in New Zealand hill country pastures. II. The fate of fertiliser sulphur. Journal of Soil Science 44: 615-624.

Smith A E and Lanfond G P 1990 Effects of long-term phenoxyalkanoic acid field applications on the rate of microbial degradation. ACS Symposium Series 426: 14-22.

Sparling G P, Murphy D V, Thompson R D and Fillery I R P 1995 Short-term net N mineralisation from plant residues and gross and net N mineralisation from soil organic N after rewetting of a seasonally dry soil. Australian Journal of Soil Research 33: 961-973.

Strong D T, Sale P W G and Helyar K R 1998 The influence of the soil matrix on nitrogen mineralisation and nitrification. I. Spatial variation and a hierarchy of soil properties. Australian Journal of Soil Research 36: 429-447.

Strong D T, Sale P W G and Helyar K R 1999 The influence of the soil matrix on nitrogen mineralisation and nitrification. IV. Texture. Australian Journal of Soil Research 37: 329-344.

Stockdale E A, Rees R M and Davies M G 1995 Nitrogen supply for organic cereal production in Scotland. *In:* Soil Management in Sustainable Agriculture. H F Cook and H C Lee (eds.) pp. 254-264. Proceeding of third international conference on Sustainable Agriculture. Wye College Press. Ashford, UK.

Thompson J P 1992 Soil biotic and biochemical factors in a long-term tillage and stubble management experiment on a Vertisol. II. Nitrogen deficiency with zero tillage and stubble retention. Soil and Tillage Research 22: 339-361.

Tilman D, Knops J, Wedin D, Reich P, Ritchie M and Siemann E 1997 The influence of functional diversity and composition on ecosystem processes. Science 277: 1300-1302.

Tlustos P, Willison T W, Baker J C, Murphy D V, Pavlikova D, Goulding K W T and Powlson D S 1998 Short-term effects of nitrogen on methane oxidation in soils. Biology and Fertility of Soils 28: 64-70.

Torsvik V, Goksøyr J, Daae F L, Sørheim R, Michalsen J and Salte K 1994 Use of DNA analysis to determine the diversity of microbial communities. *In:* Beyond the biomass. Compositional and functional analysis of soil microbial communities. K Ritz, J Dighton and K E Giller (eds.) pp. 39-48. Wiley, Chichester, New York.

Twomlow S, Riches C, O'Neill D, Brookes P and Ellis-Jones J 1999 Sustainable dryland smallholder farming in sub-saharan Africa. Annals of Arid Zone 38: 93-135.

van der Werff H M G 1996 Assessing the impact of pesticides on the environment. Agriculture Ecosystems and Environment 60: 81-96.

van Veen J A and Kuikman P J 1990 Soil structural aspects of decomposition of organic matter by micro-organisms. Biogeochemistry 11: 213-233.

Verhoef H A and Brussaard L 1990 Decomposition and nitrogen mineralisation in natural and agroecosystems: The contribution of soil animals. Biogeochemistry 11: 175-211.

Wardle D A 1992 A comparative assessment of factors which influence microbial biomass carbon and nitrogen levels in soil. Biological Reviews 67: 321-355.

Watts C W, Eich S and Dexter A R 2000 Effects of mechanical energy inputs on soil respiration at the aggregate and field scales. Soil and Tillage Research 53: 231-243.

Whitelaw M A 2000 Growth promotion of plants inoculated with phosphate-solubilising fungi. Advances in Agronomy 69: 99-151.

Willison T W, Webster C P, Goulding K W T and Powlson D S 1995 Methane oxidation in temperate soils: effects of land use and the chemical form of nitrogen fertiliser. Chemosphere 30: 539-546.

Wilkinson S C, Anderson J M, Scardelis S P, Tisiafouli M, Taylor A and Wolters V 2002 PLFA profiles of microbial communities in decomposing conifer litters subject to moisture stress. Soil Biology and Biochemistry 34: 189-200.

Wu J 1991 The turnover of organic C in soil. PhD Thesis. University of Reading. UK.

Yachi S and Loreau M 1999 Biodiversity and ecosystem productivity in a fluctuating environment. The insurance hypothesis. Proceedings of the National Academy of Sciences of the United States of America 96: 1463-1468.

Young I M and Ritz K 1998 Can there be a contemporary ecological dimension to soil biology without a habitat? Discussion. Soil Biology and Biochemistry 30: 1229-1232.

Zak J C, Willig M R, Moorhead D L and Wildman H G 1994 Functional diversity of microbial communities: a quantitative approach. Soil Biology and Biochemistry 26: 1101-1108.

Chapter 4

Role of Fauna in Soil Physical Processes

Petra C.J. vanVliet[1] and Paul F. Hendrix[2]

[1] Wageningen University, Sub-department of Soil Quality, Wageningen, The Netherlands.
[2] University of Georgia, Institute of Ecology and Department of Crop & Soil Sciences, Athens, GA, USA.

1.   INTRODUCTION

Soils are formed through the interactions of several factors: climate, parent material, organisms and topography (relief), all acting through time (Jenny 1941). Soil is the outer layer of the crust that covers the land surface of the earth and is the product of mechanical, chemical and biological weathering of parent material. For instance, in the Negev desert in Israel three species of snails feed on endolithic lichens, which grow in the limestone rocks. In order to eat the lichens, the snails must ingest rock, excreting the rock materials as faeces. Snails, at a density of $21/m^2$, convert rock to soil at a rate of ca. 70 to 110 $g/m^2/yr$ (Shachak et al. 1987, 1995).

Besides the weathering process, water and the activities of soil organisms cause movement of organic and inorganic materials in the soil profile, thereby contributing to the formation of soils. The end product is a physical mixture of inorganic particles, organic matter, air and water. Porosity, which is that portion of the soil occupied by air and water, is a very important property of soils and strongly depends on and affects abiotic and biotic conditions. Very large pores, with a diameter greater than 100 μm,

61

L.K. Abbott & D.V. Murphy (eds) Soil Biological Fertility - A Key to Sustainable Land Use in Agriculture. 61-80.
© 2007 Springer.

conduct water only during flooding or ponding rain; they are empty under drier conditions. Smaller pores almost always contain capillary (diam. 25-100 µm) and bound water (diam < 0.2 µm).

Soil texture, soil structure and porosity are interconnected and influence water transport, soil temperature, air transport and mechanical impediment of soil seedling emergence and root penetration. Soil structure cannot be measured directly and is therefore often described by size and shapes of aggregates, porosity and pore-size distribution (Koorevaar 1983). Effects of soil fauna on soil physical processes are therefore also expressed as effects on aggregation, porosity and pore-size distribution of the soil.

Soil provides a habitat for a huge array of small and large organisms. Some organisms complete their entire life cycle within the soil, while others take temporary refuge in the soil environment, which tends to be more constant than conditions above ground (Wood 1988). The effect that organisms have on the soil ecosystem depends on the number present and the time that the organisms reside in the soil. Hole (1981) classified organisms that participated in the dynamics of the soil ecosystem into the following categories: permanent, temporary, periodic, alternating, transient and accidental (see Table 1). The last three groups hardly have any influence on the properties of a soil ecosystem and will not be discussed in this chapter.

Table 1 Classification of animals in soil on basis of incidence in the soil (Hole 1981).

Category	Explanation	Representative fauna
permanent	all stages of the animal reside in the soil	Acari (mites), Collembola (springtails), earthworms
temporary	one active stage of the animal lives in the soil; another active stage does not or is periodic	larvae of many insects
periodic	the animal moves in and out of soil frequently	active forms of many insects
alternating	one or more generations of the animal live in the soil, the other generation(s) live(s) above the soil	potato aphid (*Rhopalosiphonius*), oak apple gall wasp (*Biorhiza*)
transient	inactive stages (eggs, pupae, hibernating stages) of the animal live in soil; active stages do not	many insects
accidental	the animal falls or is blown or washed into the soil	insect larvae of the forest canopy, surface animals that fall into cracks of vertisols

Soil fauna are divided into three groups according to their activities and distribution in the soil: i) epigeics that process organic matter on or near to the soil surface, ii) endogeics, which live in the mineral soil and feed on humus, and iii) anecics, which transfer materials between the soil and litter habitats (Bouché 1977). Through their location in the soil profile, these groups of animals have different effects on soil structure and physical processes.

In the first part of this chapter, we will discuss the possible effects that soil fauna can have on physical properties and processes. We conclude with an overview of the importance of soil fauna for soil physical processes in agricultural ecosystems and consider how different management strategies will influence the relationship between soil fauna and physical processes.

2. FAUNAL INFLUENCES ON SOIL PHYSICAL PROPERTIES

Activities of soil fauna that significantly affect soil structure result from the following (Lee and Foster 1991):

- burrowing and excavation in search of food, or for construction of living spaces or storage chambers within the soil or above the soil surface (e.g. earthworms, termites, ants)
- active transport of excavated or ingested soil which is deposited elsewhere (e.g. ants, earthworms)
- ingestion of soil materials (e.g. earthworms, termites)
- production of faecal pellets (e.g. microarthropods)
- use of excreta, mucus, or salivary secretions to line burrows/galleries or for gluing materials (e.g. termites, earthworms)
- collection of plant litter, animal dung, carrion from the soil surface and incorporating this into the soil with or without prior digestion (e.g. earthworms, dungbeetles).

Each of these activities has different effects on soil physical properties. Litter removal from the soil surface increases temperature gradients in the topsoil, increases evaporation of soil water, increases the possibilities for soil crusting and surface flow, and decreases infiltration (Anderson 1988). For instance, the removal of vegetation cover by grass-harvesting termites in Australia has been reported to increase sheet erosion (Wood and Sands 1978). Comminution of litter increases the active surface of the organic matter in and on the soil thereby affecting the wettability and water-holding capacity of the soil. Burrowing in the soil increases the porosity of the soil, which can have positive effects on infiltration, aeration and rooting depth of the soil. Some animals, such as earthworms, while burrowing, mix organic matter in the soil thereby increasing the water-holding capacity and aggregate stability. Certain soil fauna deposit their casts elsewhere in the soil profile or on the surface, causing a mixing of soil horizons. Casts deposited on the soil surface create a heterogenous soil surface, which limits surface flow and positively affects infiltration. However, casts can be less stable than other soil aggregates and can increase sediment erosion (Shipitalo and Protz 1988).

Figure 1 summarises the influences of soil biota on one another and on soil structure in terrestrial ecosystems. Microflora (bacteria and fungi) produce organic compounds that bind aggregates and hyphae entangle particles onto aggregates. Microfauna (protozoa and nematodes) do not have a direct effect on soil structure processes; they affect aggregate structure through their regulation of bacterial and fungal populations. The mesofauna (microarthropods and enchytraeids) affect soil structural processes through their production of faecal pellets and biopores. Some animals in this group contribute significantly to the comminution of litter material. This has a strong effect on structural stability and water-holding capacity of the soil. The macrofauna (earthworms, termites and ants) have the largest influence on soil structural processes. They create biopores and mix organic material and mineral particles. Their burrows often stimulate infiltration and aeration of the soil.

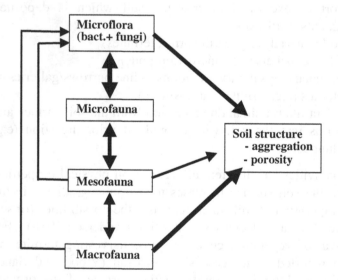

Figure 1 Direct and indirect influences of microflora, microfauna, mesofauna and macrofauna on soil structure. The magnitude of the influence is indicated by the thickness of the solid line between the boxes. Bact = bacteria

The ultimate effect of soil fauna on soil physical properties is often the result of the balance between compaction and decompaction processes (Lavelle 1997). Some invertebrates generate large aggregates that can compact the soil. Other invertebrates break these aggregates into smaller ones, by eating them and releasing small faecal pellets or by simply digging their way through these structures; this can be described as a decompaction process. In the following paragraphs the effects of several soil faunal groups on soil structure through their faecal pellets and burrowing behaviour are discussed.

2.1 Organisms which affect Soil Structure Through the Production of Faecal Pellets

Animal groups whose contribution to the formation of soil structure is limited to the production of faecal pellets are discussed. These animals hardly ingest any mineral particles and move through the soil using existing pores/burrows, thereby mixing organic matter into the topsoil. Their faecal pellets (microaggregates) may, in turn, serve as building blocks for macroaggregates (Tisdall and Oades 1982).

2.1.1 Diptera and macroarthropods

Larvae of Diptera that occur in the soil ecosystem mostly form spherical, cylindrical or spindle-like droppings, which contain large pieces of plant tissues, sometimes mixed with mineral particles. Droppings of litter-consuming diplopods and isopods contain large pieces of litter fragments, droppings of smaller soil animals or a great quantity of mineral particles, and are not very compact (size 0.5-4 mm). Because these animals occur mostly on the soil surface, their faecal pellets may accumulate to form the H-horizon of specific humus type soils (Delecour 1980). These 'organic' faecal pellets contribute to the water-holding capacity of the soil.

2.1.2 Microarthropods (mites and collembola)

Faecal pellets of oribatid mites are egg-shaped or spherical, with a smooth surface, very compact and without mineral particles inside, light brown coloured and up to 140-200 μm in size, depending on the species and life-stage (Rusek 1985). Droppings of collembola are usually compact, 30-90 μm in diameter, irregularly round, with a rugged, irregular surface, usually containing some mineral particles and usually black. Collembola have a leading role in the formation of soil microstructure in some arctic, alpine and weakly developed soils. Microarthropod faeces contribute to the water-holding capacity and aggregation of a soil.

Rusek (1985) found that some Onychiurid collembola and oribatid mites can make "microtunnels" in the soil matrix. These channels are important for capillary rise of moisture in the soil, aeration and fast drainage.

2.2   Ecosystem Engineers

Organisms that directly or indirectly modulate the availability of resources (other than themselves) to other species, thereby causing physical state changes in biotic and abiotic materials, are called 'ecosystem engineers' (Lawton and Jones 1995). These organisms modify, maintain and/or create habitats by transforming living or non-living materials from one physical state to another via mechanical or other means. Ecosystem engineers have the ability to move through soil and to build organo-mineral structures.

2.2.1   Enchytraeids

Several researchers have concluded that enchytraeids create burrows (Jegen 1920, Rusek 1985, van Vliet *et al.* 1993). Didden (1990) suggested that enchytraeids increase pore continuity and pore volume according to their body size (50-200 µm). Activities of the enchytraeids result in significant effects on air permeability, pore structure and aggregate stability. The impact of enchytraeids on hydraulic conductivity depends on the incorporation of organic matter in the soil, on the number of enchytraeids present and on the duration of the experiment (van Vliet *et al.* 1998).

Enchytraeids produce faecal pellets containing fine particles with little cellulosic plant residues. In mineral soils the faecal pellets are sponge-like structures with humus and loamy material combined (Kasprzak 1982). Didden (1990) concluded that enchytraeids have a positive effect on aggregate stability in the 600-1000 µm fraction. These aggregates might have been partly composed of enchytraeid excrements with a size of about 200 µm. However, he observed no significant effects of enchytraeids on the distribution of water-stable aggregates.

Many researchers have reported considerable amounts of mineral particles in the gut contents of enchytraeids (e.g. Babel 1968, Toutain *et al.* 1983, Didden 1990, van Vliet *et al.* 1995). Together with mineral particles attached to the body surface (Ponge 1984), enchytraeids transport these particles and may in this way influence soil structure. The amount of soil turned over by enchytraeids differs greatly between systems (Table 2). Environmental conditions, which influence enchytraeid abundances and species composition, seem to have a major influence on the soil turnover by the enchytraeid community in an ecosystem.

2.2.2   Earthworms

Numbers of earthworm burrows per unit area vary with the population density. Many researchers report about 100-300 burrows/m^2. Bouché (1971) counted more than 800 burrows/m^2 in a French pasture soil. In an irrigated orchard soil in Australia, Tisdall (1978) reported more than 2000 burrows/m². The diameter of the burrows varies with the size of the earthworm; generally the diameter of the burrows fits in the range of 1-10 mm. Due to their large size, earthworm burrows are important for fast drainage and aeration. Burrows of anecic species are open at the surface and may penetrate more than 1m deep into the soil. Each burrow is a distinct structure (Lee and Foster 1991). Burrows of endogeic species are often more or less randomly distributed through the soil, progressively decreasing in number with depth. The burrow system of endogeics changes continuously as the earthworms forage for food and fill certain sections with their casts (Lee and Foster 1991).

Table 2 An overview of soil turnover rates of several ecosystem engineers.

Ecosystem engineer	System/Country	Soil turnover rate	Reference
Enchytraeids	Arable field (The Netherlands)	0.75 t/ha/yr	Didden 1990
Enchytraeids	Arable field (USA)	21.8 t/ha/yr	van Vliet et al. 1995
Enchytraeids	Forest (USA)	4.2 t/ha/yr	van Vliet et al. 1995
Earthworms	Moist savanna (Ivory Coast)	500-1000 t/ha/yr	Lavelle 1978
Earthworms	Mexico	400 t/ha/yr	Barois et al. 1993
Earthworms	Temperate pasture	40-70 t/ha/yr	Bouché 1982
Humivorous termites	Moist savanna (Ivory Coast)	45 t/ha/yr	Lavelle et al. 1997
Termites	Guinean savanna (Niger)	0.9 t/ha/yr (litter)	Collins 1981
Termites	Senegal	2 t/ha/yr	Lepage 1974
Subterranean termites	Sonoran desert	0.75 t/ha/yr	Nutting et al. 1987
Mound building termites	Tropical Australia	0.3-0.4 t/ha/yr	Coventry et al. 1988
Ants	Subtropical regions	10 t/ha/yr	Paton et al. 1996
Ants	Desert (Australia)	0.42 t/ha/yr	Briese 1982
Ants	Argentina	2.1 t/ha/yr	Folgarait 1998
Ants	USA	0.842 t/ha/yr	Whitford et al. 1986
Tubificids	Lake sediments	2.4 t/ha/yr	Davis 1974

Many researchers have found that water infiltration increases 2 to 10 times if earthworms are present (see Lee 1985). However, this depends on the species present and environmental conditions (especially soil texture, organic matter content and climate). Burrows of anecic species might be tightly sealed by the earthworm's body; the burrow is in that case not significant for the infiltration of ponded surface water. Elimination of earthworms from a pasture resulted in a 3-fold reduction in water infiltration (Sharpley *et al.* 1979).

Inoculations of lumbricids in a Dutch polder resulted after 10 years in a significant change in soil physical properties. Uninoculated areas were covered with inactive organic mats, the mineral soil had a high penetration resistance and infiltration of water was low. In the areas with earthworms, the organic mat had disappeared, there was a high water permeability and low penetration resistance (Hoogerkamp *et al.* 1983). Twenty years of absence of earthworms in a grassland soil lead to a build-up of a litter layer and caused a compacted soil (Clements *et al.* 1991).

The size, structure and internal composition of earthworm casts depend on the ecological group that produces them. Epigeic animals usually produce cylindrical or irregular droppings containing plant material of different stages of degradation mixed with some mineral particles. These faecal pellets are not stable for a very long time (Lavelle. 1997). Burrowing (anecic) and soil feeding (endogeic) earthworms, ingest a mixture of organic and mineral debris. Casts are wet and pasty when egested, very fragile and may be easily dispersed (Shipitalo and Protz 1988). They may strongly contribute to soil loss and crust formation, especially in places where rainfall can be intense (Blanchart *et al.* 1999). In a study by Sharpley *et al.* (1979), the sediment load of surface runoff increased through splash dispersal of earthworm casts. But as the burrows of the earthworms acted as sinks for the suspended material, net sediment losses were lower in plots with earthworms than when earthworms were absent. With time and drying or drying-rewetting cycles, casts become more stable (Shipitalo and Protz 1988, Marinissen and Dexter 1990, Hindell *et al.* 1994) and become more important for aeration and drainage in soils. Dried earthworm casts are 2.5 times more resistant than similar sized dried soil aggregates (McKenzie and Dexter 1987). Stable casts increase the roughness of the soil surface, thereby reducing runoff and improving infiltration.

Earthworms in temperate pastures and grasslands deposit on average 40-50 t/ha/yr of cast material on the surface (Lee 1985). Some species in the humid tropics feed on the soil in the deeper horizons (20-50 cm) and egest casts in the soil profile; in the humid savannas of Lamto (Ivory coast), only 3 to 18% of the soil was egested as surface casts, depending on the species (Lavelle 1978). Lavelle estimated the casting rate of *Millsonia anomala* in these systems as 22-28 t/ha/yr. This equals the creation of 30-40m³/ha of pore space.

Through their foraging activities, anecic earthworms can decrease the area of soil surface protected by residual organic matter. Without residue cover, the soil surface is exposed to 'the elements' (Freebairn *et al.* 1991) and surface crusts or seals can be formed (Shuster *et al.* 2000).

Changes in the composition of the earthworm community can have far reaching effects on soil structural properties. Chauvel *et al.* (1999) showed that the invasion of the earthworm *Pontoscolex corethrurus* into soil converted from forest to pasture reduced macroporosity by 50%. Large earthworms such as *P. corethrurus* or *M. anomale* egest large and compact casts, resulting in an increase in large soil aggregates and an increase in bulk density ('compacting species'). Small earthworms such as eudrilid worms, feed at least partly on large compact casts and egest smaller, less stable aggregates, resulting in a decrease in large aggregates and a decrease in bulk density ('decompacting species') (Blanchart *et al.* 1999). The effect of the compacting species seems linked to the presence of organic residues at the surface. If organic residues are low, the presence of 'compacting' earthworms results in a lower infiltration and an increased water retention and even crust formation can occur (Blanchart *et al.* 1999, Ester and van Rozen 2002, Shuster *et al.* 2000). The introduction of decompacting species and an increase in organic residues at the soil surface will cancel these effects (Blanchart *et al.* 1999).

2.2.3   Termites

Termites are social insects living in colonies; their impact on soil structure is largely concentrated in discrete areas. Termites construct burrows to make nests, food stores or chambers for fungal gardens. They will also build vertical shafts that may penetrate several metres deep to find suitable building material or to give access to water. Horizontal burrows are constructed to provide protected access to food sources; often an extensive network of these burrows can be found (Lee and Foster 1991). The diameter of termite galleries is in the range of 1-20 mm and the network may comprise up to 7.5 km/ha (e.g. Wood 1988). Certain termite species have galleries that extend more than 50 m from the mound (Ratcliffe and Greaves 1940).

The amount of soil transported by termites is difficult to quantify because termites build a wide range of structures which are heterogeneously spaced that includes mounds (aboveground), nests (belowground), galleries and sheetings, (Anderson *et al.* 1991). Also, the soil in these structures has variable textural and chemical characteristics depending on where the soil was collected (from the soil surface or from 10-12 meters deep), if the soil was orally transported or if the soil was faecal in origin (Wood 1988). The amount of soil in surface mounds and sheetings, originating from deeper soil layers can be up to 2400 Mg/ha and cover 10% of the soil surface (Meyer 1960). Lepage (1974) found that the termite *Macrotermes subhyalinus* brought 2000 kg of soil per ha per year to the soil surface, of which between

675 and 950 kg/ha was used for the construction of foraging runways on the soil surface (Table 2). These runways rapidly erode, but are also rapidly rebuilt. Nutting *et al.* (1987) recorded a soil turnover of two species of subterranean termites of 750 kg/ha/yr (Table 2).

Soil-feeding termites ingest a mixture of organic and mineral debris. The faeces formed contain organo-mineral complexes and are stable over periods ranging from months to decades (Wood 1996). Termites generally select the smaller particles from within the soil profile and bring to the surface significant amounts of clay materials (Williams 1968, Boyer 1982). Termite mounds are often enriched with clay in comparison with unaffected soils (Lobry de Bruyn and Conacher 1990).

The effects of termite burrows on macroporosity, water infiltration and aeration are not well studied. There seem to be two contrasting theories applicable. In the first theory, termites repack the soil in such a way that it forms a compact structure which reduces water infiltration and aeration. The other theory states that termites increase infiltration and aeration by incorporating organic matter into the soil and by constructing galleries through the soil. More research is needed to determine which theory is most plausable (Lobry de Bruyn and Conacher 1990). A few studies have been done and point toward the second theory. Elkins *et al.* (1986) found that in plots in which termites were present, water infiltration rates were much higher than in plots without termites. Casenave and Valentin (1989) found that if termite sheetings covered about 30% of the soil surface infiltration increased maximally. This is probably due to increased surface roughness and the presence of galleries below the surface structures (Lavelle 1997). Others state that because the entrances to the termite galleries (burrows) are mostly closed to the surface, the effect on water infiltration is minimal (Lee and Foster 1991).

2.2.4   Ants

Ants can be found in any type of habitat in the world; they are only lacking in Iceland, Greenland, Antarctica and a few small islands (Hölldobler and Wilson 1990). Their pedobiological influence is largely through the construction of nests, galleries, soil sheetings and mounds.

As a result of the network of galleries and chambers, the porosity of a soil is increased and the soil is less compact, resulting in better drainage and aeration (Baxter and Hole 1967, Rogers 1972, Gotwald 1986, Majer *et al.* 1987, Cherrett 1989). Certain ant species fill underground cavities with a porous mixture of aggregates (750-2000 µm) (Humpreys 1994). Majer *et al.* (1987) found that the mean infiltration rate on ant nests was much faster compared with unaffected soil. However, ants expose bare soil around their burrows (Thorp 1967), which could impede water infiltration and encourage water erosion.

The movement of subterranean soil to the surface through ant activity can be very high. Paton *et al.* (1996) recorded mounding rates of 10 t/ha/yr in moist tropical and temperate ecosystems (Table 2). From a comparison of global rates of animal bioturbation, ants scored second (ca. 50 t/ha/yr) after earthworms (ca. 150 t/ha/yr), but ants have a wider geographical distribution than earthworms (Paton *et al.* 1996).

Ants also bury organic matter into the soil, thereby increasing the water-holding capacity of the soil, and mix soil layers by moving small particles from deeper layers to the surface (Petal 1978). Some researchers reported an enrichment of ant mounds with clay compared to undisturbed soil nearby, while others found the opposite (Lobry de Bruyn and Conacher 1990).

2.2.5 Tubificidae

In ricefields, tubificid worms feed with their heads downward in a burrow and deposit their faeces on the sediment surface. This leads to a vertical distribution of soil particles. Larger particles and plant residues are gradually concentrated in the lower soil layer, while finer particles are concentrated in the upper layer. Through the activities of the tubificids, the seeds of weeds are moved to a layer of about 3-5 cm below the soil surface where the oxygen concentration is too low for seed germination. Soil turnover of tubificid worms in a lake was estimated at 2.4 t/ha/yr (Table 2) (Davis 1974).

3. INFLUENCES OF MANAGEMENT ON SOIL FAUNA

The impact of soil fauna on soil structure development and stabilisation depends on the spatial and temporal scale of its actions. The following characteristics are therefore important:

- the lifetime of the individual organisms,
- the population density,
- the spatial distribution (local and regional) of the population,
- the length of time that the population has been present at the site,
- the durability of the structures in the absence of the original 'ecosystem engineers',
- the number of attributes of the ecosystem changed through the activities of the engineer (Lawton and Jones 1995).

Effects of agricultural practices on soil fauna will carry through in their influence on soil physical processes. A decrease in abundances will translate into a reduced effect of soil fauna on soil physical properties.

Table 3 shows abundances of certain soil faunal groups in different ecosystems. Most groups, which can exercise significant effects on soil structural properties, have lower abundances in arable land compared to forest and/or grassland ecosystems, depending largely on the intensity of management. Isopods are very sensitive to cultivation compared to other soil animals (Curry 1986). Their densities are often low in agricultural soils. Loss of shelter for isopods in arable land is a major factor reducing abundances (Wolters and Ekschmitt 1997). Reduced tillage operations such as minimum or no-tillage result in an increase in the biomass of isopods, compared with conventional tillage (Stinner and House 1990).

Agricultural practices (including heavy grazing, irrigation, drainage, fertilisation, mowing, ploughing) reduce ant biodiversity and or ant biomass (Folgarait 1998) and the number of colonies present. However ants can tolerate disturbances and recover and reinvade the same areas after the disturbance (Folgarait 1998). Tillage disturbs nest and channel structures of termites and certain species will disappear if they experience too much stress (Wielemaker 1984). Due to their sensitivity to disturbances, effects of termites on soil structure and soil physical processes will only be found in low-input, minimum tillage agriculture.

Table 3 Abundances (number/m^2) of different soil fauna groups in different ecosystems.

Ecosystem Soil fauna group	Forest	Grassland	Arable land	References
Isopoda	286	1200	5	Wolters and Ekschmitt 1997
Isopoda	155	nd	0	Abbott and Parker 1980
Microarthropods	208727	61885	39584	Hendrix *et al*. 1990
Enchytraeids	16200	23800	18000	Andrén and Lagerlöf 1983
Earthworms	85[a]	250	90	Paoletti 1999
Termites	2110	nd	0	Abbott and Parker 1980
Ants	155	nd	0	Abbott and Parker 1980

[a] average of deciduous, coniferous and tropical forests; nd = not determined

Soil cultivation affects soil fauna either directly or indirectly. The most obvious direct effect is the immediate killing or confinement of many epigeic and endogeic species. Indirect effects are due to repeated habitat destruction by destroying the surface layer, disturbing the natural macropores and cavities in soil, eliminating the systems of passageways created by burrowing invertebrates and, particularly in heavy soils, blocking the interstices of deeper layers by an accompanying accumulation of clay particles below the plough sole (Wolters and Ekschmitt 1997).

Ploughing or tilling reduces the earthworm population considerably (see Table 4), especially when the cultivation is followed by dry or frost periods (House and Parmelee 1985, Henke 1987). Compared to grassland, tilled areas have significantly lower numbers of burrows per area (Graff 1967, Ehlers 1975).

Soil tillage has little detrimental effect on enchytraeid abundances in contrast to earthworms, in certain systems (Table 4). There are even indications that soil tillage may positively influence enchytraeid abundances (Fründ *et al.* 1992, Didden *et al.* 1994). The smaller size of enchytraeids, their higher growth rates and less disturbance of microhabitats by tillage probably make them less vulnerable than earthworms to tillage.

Compaction of soil due to intensive traffic with heavy machinery leads to lower abundances of soil fauna (Aritajat *et al.* 1977). After slight soil compaction, recovery of the soil fauna was observed 1 month later and 3 months later when the compaction was more severe. Recovery of the soil fauna was not completed until at least 6 months after the disturbance.

Table 4 Examples of studies in which the effect of tillage on earthworm and enchytraeid abundances was determined.

Soil fauna group	Country	Conventional tillage	Minimum or no tillage	Reference
		(number/m^2)	(number/m^2)	
Earthworms	Indiana and Illinios, USA	64	138 (NT)	Kladivko *et al.* 1997
Earthworms	Georgia, USA	149	967 (NT)	Hendrix *et al.* 1986
Enchytraeids	Georgia, USA	15270	16830 (NT)	van Vliet *et al.* 1995
Enchytraeids	The Netherlands	23845	43007 (MT)	Didden 1991

*NT = no tillage; MT = minimum tillage.

Sustainable agriculture aims to reduce external inputs, enhance internal recycling of nutrients by reducing tillage, and use organic instead of mineral fertilisers (Doran and Werner 1990). Agricultural production systems emphasising minimum or zero tillage or direct drilling obviously promote earthworm populations in arable land (Table 4). These reduced soil cultivation techniques compared to conventional cultivation systems increase population densities by a factor of 2-3 (Makeschin 1997). Deep burrowing species such as *Lumbricus terrestris*, especially benefit from reduced cultivation (Edwards and Lofty 1982, Lavelle 1997). The burrows made by these species are very important for aeration and drainage during and after heavy rainfall. A few endogeic species benefit from cultivation and mixing of organic residues in arable soils. The availability of sufficient organic food seems to override the influence of cultivation practices (Doube *et al.* 1994).

In an agroecosystem with reduced tillage, more favourable habitats are created and a greater soil faunal diversity develops. Higher densities of soil fauna, and especially of the soil ecosystem engineers, result in a larger influence of these animals on soil physical processes. Conversely, reduced densities of these organisms, as a result of certain management practices, can be expected to reduce their influences on soil processes and possibly cause longer-term changes in soil physical properties. The best documented effects have been observed following reductions in earthworm population densities (e.g. Parmelee *et al.* 1990, Clements *et al.* 1991), which often result in decreased organic matter decomposition, thatch or litter accumulation, and possibly soil compaction. Similar or perhaps more subtle effects may result from population declines in other soil faunal groups (e.g. Microarthropods: Crossley *et al.* 1989) but these smaller animals may operate as engineers at a much finer scale; further research is needed to assess their influences on soil physical processes in response to management.

4. CONCLUSIONS

In this chapter, the potential contribution of several soil animal groups to soil physical processes has been described. Through their feeding, excrements and burrowing activities the influence of soil biota on soil structural properties can be substantial. Soil ecosystem engineers, which burrow through a soil, can contribute significantly to the better functioning of physical processes. However, for these influences to be significant, abundances of the different soil fauna need to be much larger than they are currently found in conventional tillage systems. The less ecosystem engineers are disturbed by a particular tillage system, the greater their influence on physical processes.

Sustainable agriculture, which reduces the use of non-renewable resources and protects the environment (Pankhurst and Lynch 1994), stimulates many biological soil processes. In low-input, minimum tillage or no-tillage systems, soil biota have a more important role in the maintenance and development of soil structure and soil physical processes, contributing to agricultural sustainability.

5. ACKNOWLEDGEMENTS

The authors are grateful to the anonymous reviewers for helpful comments. This work was supported by grants from the Netherlands Organization for Scientific Research (NWO) to Wageningen University and from the U.S. National Science Foundation to the University of Georgia Research Foundation.

6. REFERENCES

Abbott I and Parker C A 1980 Agriculture and the abundance of soil animals in the Western Australian wheatbelt. Soil Biology and Biochemistry 12: 455-59.

Anderson J, Knight D and Elliott P 1991 Effects on invertebrates on soil properties and processes. In: Advances in Management and Conservation of Soil Fauna. G K Veeresh, D Rajagopal and C A Viraktamath (eds.) pp. 473-484. Oxford and IBH Publishing Co. New Delhi.

Anderson J M 1988 Invertebrate-mediated transport processes in soils. Agriculture Ecosystems and Environment 24: 5-19.

Andrén O and Lagerlöf J 1983 Soil fauna (microarthropods, enchytraeidae, nematodes) in Swedish agricultural cropping systems. Acta Agriculturae Scandinavica 33: 33-52.

Aritajat U, Madge D S and Gooderham P T 1977 The effects of compaction of agricultural soils on soil fauna. I. Field investigations. Pedobiologia 17: 262-82.

Babel U 1968 Enchytraeen-losungsgefuge in löss. Geoderma 2: 57-63.

Barois I, Villemin G, Lavelle P and Toutain F 1993 Transformation of the soil structure through Pontoscolex corethrurus (Oligochaeta) intestinal tract. Geoderma 56: 57-66.

Baxter P F and Hole H 1967 Ant (Formica cinerea) pedoturbation in a prairie soil. Soil Science Society of America Proceedings 31: 425-28.

Blanchart E, Albrecht A, Alegre J, Duboisset A, Gilot C, Pashanasi B, Lavelle P and Brussaard L 1999 Effects of earthworms on soil structure and physical properties. In: Earthworm management in tropical agroecosystems. P Lavelle, L Brussaard and P Hendrix (eds.) pp. 149-172. CAB International. Wallingford, UK.

Bouché M B 1982 Ecosystème prairal. 4.3. Un example d'activité animale: le rôle des lombriciens. Acta Oecol.; Oecolo. Gen. 3: 127-54.

Bouché M B 1971 Relations entre les structures spatiales et fonctionelles des écosystèmes, illustrées par le rôle pédobiologique des vers de terre. In: La Vie dans les Sols. C Delamere Deboutteville (ed.) pp. 187-209. Gauthiers Villars. Paris.

Bouché M B 1977 Stratégies lombriciennes. In:" Soil Organisms as Components of Ecosystems. U Lohm and T Persson (eds.) Ecol. Bull. 25: 122-132.

Boyer P 1982 Quelques aspects de l'action des termites sur les argiles. Clay Mineral 17: 453-62.

Briese D T 1982 The effects of ants on the soil of a semi-arid saltbush habitat. Ins. Soc. 29: 375-82.

Casenave A and Valentin C 1989 Les états de Surface de la Zone Sahélienne. Influence sur l'Infiltration. pp. 229. ORSTOM. Paris, France.

Chauvel A, Grimaldi M, Barros E, Blachart E, Desjardins T, Sarrazin M and Lavelle P 1999 Pasture damage by an Amazonian earthworm. Nature 398: 32-33.

Cherrett J M 1989 Leaf-cutting ants - biogeographical and ecological studies. *In:* Ecosystems of the World - Tropical Rain Forest Ecosystem. H Leith and M J Werger (eds.) pp. 473-488. Elsevier. New York.

Clements R O, Murray P J, Sturdy R G 1991 The impact of 20 years' absence of earthworms and three levels of N fertilizer on a grassland soil environment. Agriculture, Ecosystems and Environment 36: 75-85.

Collins N M 1981 The role of termites in the decomposition of wood and leaf litter in the southern guinean savanna of Nigeria. Oecologia 51: 389-399.

Coventry R J, Holt J A and Sinclair D F 1988 Nutrient cycling by mound-building termites in low-fertility soils of semi-arid tropical Australia. Australian Journal of Soil Research 26: 375-90.

Crossley D A Jr, Coleman D C and Hendrix P F 1989 The importance of the fauna in agricultural soils: research approaches and perspectives. Agriculture, Ecosystems and Environment 27: 47-55.

Curry J P 1986 Effects of management on soil decomposers and decomposition processes in grassland. *In:* Developments in Biogeochemistry - Microfloral and Faunal Interactions in Natural and Agro-ecosystems. M J Mitchell and J P Nakas (eds) pp. 349-398. Nijhoff/W. Junk. Dordrecht, The Netherlands.

Davis R B 1974 Stratigraphic effects of tubificids in profundal lake sediments. Limnol. Oceanogr. 19: 466-88.

Delecour F 1980 Essai de classification pratique des humus. Pedologie 30: 225-41.

Didden W A M 1990 Involvement of Enchytraeidae (Oligochaeta) in soil structure evolution in agricultural fields. Biological Fertility of Soils 9: 152-58.

Didden, W A M 1991 Population ecology and functioning of Enchytraeidae in some arable farming systems. Doctoral Thesis, Agricultural University. p. 116. Wageningen, The Netherlands.

Didden W A M, Marinissen J C Y, Vreeken-Buijs M J, Burgers S L G E, Fluiter R D, Geurs M and Brussaard L 1994 Soil meso- and macrofauna in two agricultural systems: factors affecting population dynamics and evaluation of their role in carbon and nitrogen dynamics. Agriculture, Ecosystems and Environment 51: 171-86.

Doran J W and Werner M R 1990 Management and soil biology. *In:* Sustainable Agriculture in Temperate Zones. C A Francis, C B Flora and L D King (eds.) pp. 205-230. Wiley and Sons Inc. New York.

Doube B M, Buckerfield J C, Kirkegaard K A 1994 Short-term effects of tillage and stubble management on earthworm populations in cropping systems in southern New South Wales. Australian Journal of Agricultural Research 45: 1587-1600.

Edwards C A, Lofty J R 1982 The effect of direct drilling and minimal cultivation on earthworm populations. Journal of Applied Ecology 19: 723-734.

Ehlers W 1975 Observations of earthworm channels and infiltration on tilled and untilled loess soil. Soil Science 119: 242-49.

Elkins N Z, Sabol G V, Ward T J and Whitford W G 1986 The influence of subterranean termites on the hydrological characteristics of a Chihuahuan desert ecosystem. Oecologia 68: 521-28.

Ester A and van Rozen K 2002 Earthworms (*Aporrectodea* spp.; Lumbricidae) cause soil structure problems in young Dutch polders. European Journal of Soil Biology 38: 181-85.

Folgarait P J 1998 Ant biodiversity and its relationship to ecosystem functioning: a review. Biodiversity Conservation 7: 1221-44.

Freebairn D M, Gupta S C and Rawls W J 1991 Influence of aggregate size and microrelief on development of surface soil crusts. Soil Science Society of America Journal 55: 188-95.

Fründ H C, Necker U and Kamann T 1992 Bodenbiologische Untersuchungen auf Konvenionell and Organisch-Biologisch bewirtschafteten Ackerflächen. 2. Erhebungen zur Bodenfauna. VDLUFA-Schriftenreihe 35: 567-570.

Gotwald W H 1986 The beneficial economic role of ants. *In:* Economics Impact and Control of Social Insects. S B Vinson (ed.) pp. 290-313. Praeger Special Studies. New York.

Graff O 1967 Über die Verlagerung von Nährelementen in den Unterboden durch Regenwurmtätigkeit. Landw. Forschg. 20: 117-27.

Hendrix P F, Crossley D A Jr, Blair J M and Coleman D C 1990 Soil Biota as components of sustainable agroecosystems. *In:* Sustainable Agricultural Systems. C A Edwards, R Lal, P Madden, R H Miller and G House (eds.) pp. 637-654. Ankeny: Soil and Water Conservation Society.

Hendrix P F, Parmelee R W, Crossley D A Jr, Coleman D C, Odum E P and Groffman P M 1986 Detritus food webs in conventional and no-tillage agroecosystems. Bioscience 36: 374-80.

Henke W 1987 Einfluss der Bodenbearbeitung auf die Regenwumaktivität. Mitteilgn. Dtsch. Bodenkundl. Gesellsch. 55: 885-890.

Hindell R P, McKenzie B M, Tisdall J M 1994 Relationships between casts of geophagous earthworms (Lumbricidae, Oligochaeta) and matric potential. II. Clay dispersion from casts. Biology and Fertility of Soils 18: 127-31.

Hole F D 1981 Effects of animals on soil. Geoderma 25: 75-112.

Hölldobler B and Wilson E O 1990 The Ants. p. 732. Springer. Berlin.

Hoogerkamp M, Rogaar H and Eijsackers H J P 1983 Effect of earthworms on grassland reclaimed polder soils in the Netherlands. *In:* Earthworm Ecology. J E Satchell (ed.) pp. 85-105. Chapman and Hall. London.

House G J and Parmelee R W 1985 Comparison of soil arthropods and earthworms from conventional and no-tillage agroecosystems. Soil Tillage Research 5: 351-60.

Humphreys G S 1994 Bioturbation, biofabrics and the biomantle: an example from the Sydney Basin. *In:* Soil Micromorphology: Studies in management and genesis. A J Ringrose-Voase and G S Humphreys (eds.) pp. 421-436. Developments in Soil Science no. 22. Elsevier. Amsterdam, The Netherlands.

Jegen G 1920 Die Bedeutung der Enchytraeiden für die Humusbildung. Landwirtsch. Jahrb. Schweiz 34: 55-71.

Jenny H F 1941 Factors of Soil Formation. p. 281. McGraw-Hill. New York.

Kasprzak K 1982 Review of enchytraeid (Oligochaeta, Enchytraeidae) community structure and function in agricultural ecosystems. Pedobiologia 23: 217-32.

Kladivko E J, Akhouri N M and Weesies G 1997 Earthworm populations and species distributions under no-till and conventional tillage in Indiana and Illinois. Soil Biology and Biochemistry 29: 613-15.

Koorevaar P, Menelik G and Dirksen C 1983 Elements of Soil Physics. p. 228. Elsevier. Amsterdam.

Lavelle P 1997 Faunal activities and soil processes: adaptive strategies that determine ecosystem function. Advances in Ecological Research 27: 93-132.

Lavelle P 1978 Les vers de terre de la savane de Lamto (Ivory Coast): peuplements, populations et fonctions dans l'écosystème. Thèse de Doctorat. p. 310. Universite Paris VI.

Lavelle P, Bignell D, Lepage M, Wolters V, Roger P, Ineson P, Heal O W and
    Dhillion S 1997 Soil function in a changing world: the role of invertebrate
    ecosystem engineers. European Journal of Soil Biology 33: 159-93.
Lawton J H and Jones C G 1995 Linking species and ecosystems: organisms as
    ecosystem engineers. *In:* Linking Species and Ecosystems. C G Jones and J H
    Lawton (eds.) pp. 141-150. Chapman and Hall. New York.
Lee K E 1985 Earthworms: their ecology and relationships with soils and land use. pp.
    411. Sydney. Academic Press.
Lee K E and Foster R C 1991 Soil fauna and soil structure. Australian Journal of Soil
    Research 29: 745-75.
Lepage M 1974 Les termites d'une savane sahélienne (Ferlo Septentrional, Sénégal):
    peuplements, consommation, role dans l'écosysteme. Ph.D. Thesis. Univ. of Dijon.
    Dijon.
Lobry de Bruyn L A and Conacher A J 1990 The role of termites and ants in soil
    modification: a review. Australian Journal of Soil Research 28: 55-93.
Majer J D, Walker T C and Berlandier F 1987 The role of ants in degraded soils
    within Dryandra state forest. Mulga Research Centre Journal 9: 15-16.
Makeschin F 1997 Earthworms (Lumbricidae: Oligochaeta): Important promoters of soil
    development and soil fertility. *In:* Fauna in soil ecosystems: recycling processes,
    nutrient fluxes and agricultural production. G Benckiser (ed.) pp. 173-223. Marcel
    Dekker Inc. New York.
Marinissen J C Y and Dexter A R 1990 Mechanisms of stabilization of earthworm
    casts and artificial casts. Biology and Fertility of Soils 9: 163-67.
McKenzie B M and Dexter A R 1987 Physical properties of casts of the earthworm
    Aporrectodea rosea. Biology and Fertility of Soils 5: 152-57.
Meyer J A 1960 Résultats agronomiques d'un essai de nivellement des termitières
    réalisé dans la cuvette centrale Congolese. Bull. d'Agric. Congo Belge 51: 1047-
    1059.
Nutting W L, Haverty M I and La Fage J P 1987 Physical and chemical alteration of
    soil by two subterranean termite species in Sonoran Desert grassland. Journal of Arid
    Environments 12: 233-39.
Pankhurst C E, Lynch J M 1994 The role of soil biota in sustainable agriculture. *In:*
    Soil Biota: Management in sustainable farming systems. C E Pankhurst, B M Doube,
    V V S R Gupta and P R Grace (eds.) pp. 3-9. CSIRO. Melbourne, Australia.
Paoletti M G 1999 The role of earthworms for assessment of sustainability and as
    bioindicators. Agricriculture Ecosystems and Environment 74: 137-55.
Parmelee R W, Beare M H, Cheng W X, Hendrix P F, Rider S J, Crossley, D A Jr
    and Coleman D C 1990 Earthworms and enchytraeids in conventional and no-
    tillage agroecosystems: A biocide approach to assess their role in organic matter
    breakdown. Biology and Fertility of Soils 10: 1-10.
Paton T R, Humphreys G S and Mitchell P B 1996 Soils: a new global view. p. 234.
    Yale University Press. New Haven.
Petal J 1978 The role of ants in ecosystems. *In:* Production ecology of ants and
    termites. M V Brain (ed.) pp. 293-325. Cambridge Univ. Press. Cambridge, UK.
Ponge J F 1984 Étude écologique d'un humus forestier par l'observation d'un petit
    volume, premiers résultats. I. La couche L1 d'un moder sous pin sylvestre. Rev.
    Écol. Biol. Sol. 21: 161-87.
Ratcliffe F N and Greaves T 1940 The subterranean foraging galleries of *Coptotermes
    lacteus* (Frogg.). J. Counc. Sic. Ind. Res. 13: 150-61.

Rogers L E 1972 The ecological effects of the western harvester ant (*Pogonomyrmex occidentalis*) in the shortgrass plains ecosystem. Grassland Biome. Technical Report. pp. 206. USA/IBP.

Rusek J 1985 Soil microstructures - contributions on specific organisms. Quaest. Entomol. 21: 497-514.

Shachak M, Jones C G and Brand S 1995 The role of animals in an arid ecosystem: snails and isopods as controllers of soil formation, erosion and desalinization. Advances in GeoEcology 28: 37-50.

Shachak M, Jones C G and Granot Y 1987 Herbivory in rocks and weathering of a desert. Science 236: 1098-99.

Sharpley A N, Syers J K, and Springett J A 1979 Effect of surface-casting earthworms on the transport of phosphorus and nitrogen in surface runoff from pasture. Soil Biology and Biochemistry 11: 459-62.

Shipitalo M J and Protz R 1988 Factors influencing the dispersibility of clay in worm casts. Soil Science Society of America Journal 52: 764-69.

Shuster W D, Subler S and McCoy E L 2000 Foraging of deep-burrowing earthworms degrades surface soil structure of a fluventic Hapludoll in Ohio. Soil Tillage Research 54: 179-89.

Stinner B R and House G J 1990. Arthropods and other invertebrates in conservation-tillage agriculture. Annual Review of Entomology 35: 299-18.

Thorp J 1967 Effects of certain animals that live in soils. *In:* Selected papers in soil formation and classification. SSSA Special Publication Series, Nr. 1. J V Drew (ed.) pp. 191-208. Soil Science Society of America, Inc. Madison, Wisconsin, USA.

Tisdall J M 1978 Ecology of earthworms in irrigated orchards. *In:* Modification of soil structure. W W Emerson, R D Bond and A R Dexter (eds.) pp. 297-303. Wiley and Sons. Chichester, UK.

Tisdall J M and Oades J M 1982 Organic matter and water-stable aggregates in soils. Journal of Soil Science 33: 141-63.

Toutain F, Brun J J and Rafidison Z 1983 Role des organismes vivants dans les arrangements structuraux des sols: Biostructures et mode d'alteration. Sci. Geol. Mem. 73: 115-22.

van Vliet P C J, Beare M H and Coleman D C 1995 Population dynamics and functional roles of Enchytraeidae (Oligochaeta) in hardwood forest and agricultural ecosystems. *In:* The Significance and Regulation of Soil Biodiversity. H Collins, G P Robertson and M J Klug (eds.) pp. 237-245. Kluwer Academics. Dordrecht, The Netherlands.

van Vliet P C J, Ratcliffe D E, Hendrix P F and Coleman D C 1998 Hydraulic conductivity and pore-size distribution in small microcosms with and without enchytraeids (Oligochaeta). Applied Soil Ecology 9: 277-82.

van Vliet P C J, West L T, Hendrix P F and Coleman D C 1993 The influence of Enchytraeidae (Oligochaeta) on the soil porosity of small microcosms. Geoderma 56: 287-99.

Whitford W G, Schaeffer D and Wisdom W 1986 Soil movement by desert ants. Southwest. Nat. 31: 273-74.

Wielemaker W G 1984 Soil formation by termites; a study in the Kisii Area, Kenya. Doctorate Thesis. Wageningen University. pp. 132. The Netherlands.

Williams M A J 1968 Termites and soil development near Brooks Creek, Northern Territory. Australian Journal of Soil Science 31: 153-54.

Wolters V and Ekschmitt K 1997 Gastropods, isopods, diplopods and chilopods: Neglected groups of the decomposer food web. *In:* Fauna in soil ecosystems:

recycling processes, nutrient fluxes and agricultural production. G Benckiser (ed.)
pp. 265-306. Marcel Dekker Inc. New York.

Wood T G 1988 Termites and the soil environment. Biology and Fertility of Soils 6:
228-36.

Wood T G 1996 The agricultural importance of termites in the tropics. Agr. Zool. Rev.
7: 117-55.

Wood T G and Sands W A 1978 The role of termites in ecosystems. *In:* Production
Ecology of Ants and Termites. M V Brian (ed.) pp. 245-292. University Press.
Cambridge.

Chapter 5

Contributions of Rhizosphere Interactions to Soil Biological Fertility

Petra Marschner[1] and Zdenko Rengel[2]

[1] Soil and Land Systems, School of Earth and Environmental Sciences, The University of Adelaide, Waite Campus, PMB1, Glen Osmond SA, 5064, Australia.
[2] School of Earth and Geographical Sciences, Faculty of Natural and Agricultural Sciences, The University of Western Australia, Crawley, 6009, WA, Australia.

1. INTRODUCTION

Availability of nutrients in the rhizosphere, which is defined as the soil around the root that is influenced by the root, is controlled by the combined effects of soil properties, plant characteristics, and the interactions of plant roots with microorganisms and the surrounding soil (Bowen and Rovira 1992). The rhizosphere extends up to a few millimetres from the root surface into the surrounding soil and is characterised by a high concentration of easily degradable substrates in root exudates (Lynch and Whipps 1990), which leads to a proliferation of microorganisms (Foster 1986, Curl and Truelove 1986, Rouatt and Katznelson 1961) (Table 1). Root exudation is greatest at the root tip (Marschner 1995) where the microbial density is low (Schönwitz and Ziegler 1989). With increasing distance from the root tip, exudation generally decreases while microbial density increases. Thus, the region of greatest release of root exudate and the region of highest microbial population density are spatially separated. Compared to the bulk soil, nutrient concentrations in the rhizosphere may be increased or decreased

L.K. Abbott & D.V. Murphy (eds) Soil Biological Fertility - A Key to Sustainable Land Use in Agriculture. 81-98.

(Hendriks *et al*. 1981) (Table 1). The rhizosphere pH is increased by plant nitrate uptake while ammonium uptake leads to a pH decrease. A pH decrease is also observed in the rhizosphere of N_2 fixing legumes (Römheld 1986) (Table 1).

Table 1 Characteristics of the rhizosphere compared to the bulk soil.

Compared to the bulk soil the rhizosphere is characterised by...

Higher concentrations of sugars, organic acids, amino acids and other easily degradable substances

Higher density of microorganisms

Increased or decreased nutrient concentration

Increased or decreased pH

Besides serving as a carbon source for microorganisms, root exudates also play an important role in nutrient release by chelation and desorption of poorly soluble nutrients such as phosphorus (P) and iron (Fe) (Dinkelaker and Marschner 1992, Gerke 1994, Jones and Darrah 1994, Römheld 1991, Uren and Reisenauer 1988). For example, increasing amounts of citrate or oxalate adsorbed to the soil matrix result in increasing P mobilisation by ligand exchange and dissolution of P-sorbing Fe and aluminium (Al) sites (Gerke *et al*. 2000). Some exudates, such as amino acids, are even taken up again by the roots, thus minimising nitrogen (N) loss (Jones and Darrah 1994).

Despite the importance of rhizosphere processes in influencing nutrient availability, until recently these processes had not received major consideration in modern agriculture, where the practice has been to provide N, P, and potassium (K) in luxurious quantities as synthetic fertilisers (Schaffert 1994). This has generated concern that the selection and breeding of new plant genotypes for agriculture has resulted in the development of cultivars that are highly responsive to fertilisers, but that do not have traits that are necessary for growth under nutrient-limiting or adverse soil conditions (Duncan and Baligar 1990). With the current emphasis on developing better cultivars for sustainable, low-input agriculture, a better knowledge of the rhizosphere processes that contribute to nutrient uptake efficiency has become essential for nutrient-limiting soils (Crowley and Rengel 1999).

In this chapter, the importance of the interactions in the rhizosphere for soil biological fertility will be discussed with respect to root exudates, P, Fe, manganese (Mn) and plant growth-promoting rhizosphere organisms.

2. ROOT EXUDATES AND MICROORGANISMS

Compared to the soil organic matter, root exudates represent an easily degradable nutrient source for microorganisms, and some microbial species proliferate rapidly in the rhizosphere. The microbial biomass may comprise up to 36% of root dry weight (Whipps and Lynch 1983). These are usually species with high growth rates and relatively high nutrient requirements, such as pseudomonads (Marilley and Aragno 1999). Microorganisms such as N_2 fixers are also attracted by signalling substances excreted by roots (Sheng and Citovsky 1996, Zhu *et al.* 1997). On the other hand, microorganisms can enhance the release of root exudates (Meharg and Kilham 1995) and produce growth factors that influence root growth (Arshad and Frankenberger 1993).

Of the carbon assimilated by the plant, 60-80% remains in the plant, 10-25% is used in root respiration and 11-18% is released into the rhizosphere (Warembourg and Billes 1973, Haller and Stolp 1985, Merbach *et al.* 1999). Depending on the plant species, the amount of C released into the rhizosphere ranges from 0.1 to 1.5 t C per ha (Merbach *et al.* 1999). From 40 to 80% of this carbon is respired by microorganisms, leaving 40-600 kg C per ha that is incorporated into microbial biomass or is broken down chemically (Haller and Stolp 1985, Merbach *et al.* 1999).

Plant species differ in composition and amount of root exudates they produce (Vancura and Hovadik 1965, Merbach and Ruppel 1992) and in their composition of rhizosphere microflora (Lemanceau *et al.* 1995, Wiehe and Höflich 1995, Martin 1971, Grayston *et al.* 1998, Marschner *et al.* 2001). Even genotypes within a species may have a specific rhizosphere microflora (Timonin 1946, Howie and Echandi 1983, Rengel *et al.* 1996). Additionally, the microbial community composition depends on the plant growth stage, P and N fertilisation and other soil factors (Brown *et al.* 1973, Jiang and Sato 1992, 1994, Aulakh *et al.* 2001).

By utilising root exudates for respiration, growth and metabolism, microorganisms in the rhizosphere may decrease nutrient mobilisation by root exudates. The half-life of organic acids and amino acids in soil is 2-3 h and 12 h, respectively (Jones 1999, Jones and Darrah 1994), as they are rapidly degraded by soil microorganisms. However, organic acids, such as citrate, may be strongly sorbed onto soil components, a process that reduces organic acid decomposition by soil microorganisms. Within 10 minutes after addition, 99% and 83% of the added citrate was adsorbed to Fe oxides and kaolinite, respectively, this reduced citrate decomposition by 99% in Fe oxides and 75% in kaolinite (Jones and Edwards 1998) (Figure 1).

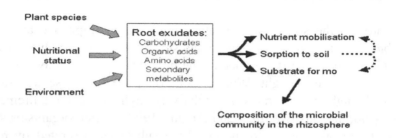

Figure 1 Role of exudates for rhizosphere processes. mo = microorganisms

3. AVAILABILITY OF PHOSPHORUS IN THE RHIZOSPHERE

Although the total amount of P in the soil may be high, it is mainly present in forms unavailable to plants. The accessibility of applied P by crop plants is often very low because more than 80% of added P becomes immobile and poorly available for plant uptake due to adsorption, precipitation, or conversion to the organic forms (Schachtman *et al*. 1998). Organic P, predominantly phytic acid, may represent up to 80% of total soil P. In alkaline calcareous soils, P is often precipitated as calcium (Ca) phosphate. Acidification of the rhizosphere by extrusion of protons causes dissolution of plant-unavailable P forms such as rock phosphate (Trolove *et al*. 1996) and Ca-P complexes in calcareous soils (Yan *et al*. 1996). In acidic soils P can be present as Fe/Al phosphates, adsorbed to Fe/Al oxides or humic substances.

As a response to the low P availability, organisms have evolved different strategies to increase P uptake. Under P deficiency, plants may increase root growth, form thinner roots or increase root hair length (Föhse and Jungk 1983) thereby increasing root surface area. Plants and microorganisms also release organic acids such as citrate, malate and oxalate, which increase solubility of inorganic P by ligand exchange and dissolution of P-sorbing Fe and Al sites (Banik and Dey 1983, Gerke 1994, Hoffland *et al*. 1989, Neumann *et al*. 1999). Plant species and genotypes within species that differ in tolerance to P deficiency also differ in solubilising activity of their root exudates (Caradus 1995, Subbarao *et al.* 1997). It is generally accepted that the type of exudates plants release into the rhizosphere plays a significant role in distribution of plant species in ecosystems. Calcicole plants exude mostly di- and tri-carboxylic acids (high mobilisation capacity for P, Fe and Mn from calcareous soils), while calcifuge plants exude mostly monocarboxylic acids (poor in mobilising P or Fe from calcareous soils) (Tyler and Ström 1995, Ström 1997). Similar adaptations are found in crop plants too (cf. Zhang *et al.* 1997) (Figure 2). By selecting genotypes with a high capacity to excrete organic acids poorly available P sources can be utilised. This P solubilisation may even benefit neighbouring plants. If the crop residues are used as mulch, P taken up by the effective genotype can also become available for the following crop.

A large number of microorganisms have shown the capacity to solubilise sparingly soluble P *in vitro* (e.g. Banik and Dey 1983, Whitelaw *et al.* 1999) (Figure 2). It has been proposed that inoculation with effective microbial P solubilisers may increase P uptake and growth of plants. Indeed, in several pot or field experiments increased P uptake and growth after inoculation with P-solubilising microorganisms have been reported (Gerretsen 1946, Kumar and Narula 1999, Kundu and Gaur 1980). Gerretsen (1946) for example, showed that compared to the sterile plants, P uptake by rye, oats, mustard and canola from poorly soluble P sources was increased by inoculation with soil microorganisms. However, it is often unclear whether the improved P uptake of the plants is directly due to P solubilisation by the introduced microbe. Often a combination of microorganisms with different characteristics, such as P solubilisers combined with N_2 fixers or with AM fungi, is superior to inoculation with the P solubiliser alone (Kumar and Narula 1999, Toro *et. al.* 1997).

Despite these successes, large-scale inoculation with P solubilisers is hampered by several factors that may decrease the effect of the introduced microorganisms:

- most soils already contain P solubilisers so that the effect of an additional strain may be small,
- the survival of the introduced strain in the rhizosphere may be low because it has a low competitiveness against indigenous, well-adapted species,
- microorganisms are selected based on their P solubilisation *in vitro*: however, the selective media are usually not buffered and have high concentration of nutrients other than P, thus presenting ideal conditions for growth and P solubilisation. In the rhizosphere, P solubilisation may be much lower as other nutrients limit growth and production of organic acids, and

- P solubilised by the microorganisms may not be available for the plant as it is rapidly taken up by the microorganisms themselves.

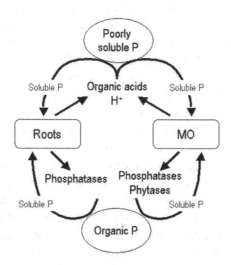

Figure 2 Phosphorus availability in the rhizosphere. mo = microorganisms

The conditions for an effective interaction between P solubilisers and plants are:

- high population of the P solubiliser maintained over long periods in the rhizosphere,
- exudation of organic acids and protons in the rhizosphere by roots and microorganisms,
- low P uptake by microorganisms, and
- a positive interaction with mycorrhizal fungi or other beneficial microorganisms.

Although a large proportion of total P is present as organic P, little is known about possible stimulation of plant P uptake from this source by microorganisms (Richardson 2001). In both plants and microorganisms, the activity of P-hydrolysing enzymes is increased under P deficiency. These enzymes break down organic P, thus making P available for uptake. They are grouped according to the compound they hydrolyse (phosphatases, phosphodiesterase and phytase) and their pH optima. Plant roots and microorganisms produce acid phosphatases that show the highest activity at pH 6. Alkaline phosphatases with a pH optimum of 11 are produced only by microorganisms.

Exudation of phosphatases occurs from the apical root region and increases under P-deficient conditions in a number of crop species (Tarafdar and Jungk 1987). The activity of phosphatases decreases with the distance from the root surface (Li *et al.* 1997-b, Tarafdar and Jungk 1987).

Phosphatases are not effective in decomposing phytin (inositol hexaphosphate), which represents the majority of stored organic P in many soils. Instead, the enzyme phytase specifically catalyses the break-down of phytin. Plant roots do not exude phytases (Li *et al.* 1997-a), it is only through microorganisms (e.g. *Aspergillus niger*) excreting phytases that phytin may become a source of P to plants (see Delhaize 1995, Richardson *et al.* 2000). Mycorrhizal fungi, the symbiotic organisms that play a key role in P nutrition of plants, are discussed in Chapter 7.

4.   INTERACTIONS BETWEEN PLANT ROOTS AND MICROORGANISMS FOR IRON

Similar to P, the total Fe content in soil is relatively high but its availability to plants is low in aerated soils because the prevalent form (Fe^{3+}) is poorly soluble. Plants and microorganisms have developed mechanisms to increase Fe uptake. In plants, two different strategies in response to Fe deficiency are evident.

Strategy I plants (dicots and non-graminaceous monocots) release organic acids that chelate Fe. Iron solubility is increased by decreasing the rhizosphere pH, and Fe uptake is enhanced by an increased reducing capacity of roots ($Fe^{3+} \rightarrow Fe^{2+}$). Additionally, root morphology and histology may change (root tip swelling, increased root branching, more root hairs, formation of rhizodermal transfer cells, etc.).

Strategy II plants (grasses) release phytosiderophores (PS) that chelate Fe^{3+}. PS are non-proteinogenic amino acids (Takagi *et al.* 1984) with high specificity for certain metals. Iron is taken up in the chelated form as Fe-PS by a specific uptake system that is strongly activated under Fe deficiency (Römheld 1991, Von Wiren *et al.* 1994). They are released only a few hours a day in a specific diurnal pattern at the root tip (Römheld 1986). The rate of release of PS is positively related to Fe-efficiency of species. The relative effectiveness in PS release, and thus Fe efficiency, decreases in the order: barley > corn > sorghum (Römheld and Marschner 1990) or oats > corn (see Mori 1994). Thus by selecting Fe efficient crops such as barley and oats, the need for Fe fertilisation can be reduced. When used as ground cover, Fe efficient grasses can even improve Fe uptake by plant species susceptible to Fe deficiency such as guava (Kamal *et al.* 2000).

Microorganisms can also release organic acids, but under Fe deficiency many produce siderophores that chelate Fe^{3+}. After translocation of the ferrated chelate, Fe is released by reduction either outside or within the cell (Neilands 1984). A range of siderophores are found in microorganisms, e.g.

ferrichromes in fungi, and enterobactin, pyoverdine and ferrioxamines in bacteria. A given species may produce one or several siderophores and have the capacity to take up not only its own siderophores but also those of other species (Raaijmakers *et al.* 1995).

Bacterial siderophores were often found to be poor Fe sources for both monocot and dicot plants (Bar-Ness *et. al.* 1992, Walter *et al.* 1994, Crowley *et al.* 1992). However, in some cases bacterial siderophores prevented Fe deficiency-induced chlorosis in dicots (Jurkevitch *et al.* 1988, Wang *et al.* 1993). Iron chelated by rhizoferrin from the fungus *Rhizopus arrhizus* was a good Fe source for barley (Yehuda *et al.* 1996). On the other hand, Fe-PS complexes appear to be a good Fe source for bacteria (Jurkevitch *et al.* 1993, Marschner and Crowley 1998) (Figure 3).

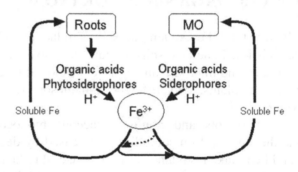

Figure 3 Interactions between roots and rhizosphere microorganisms (MO) for iron.

The interactions between different Fe chelators are complex; they are a function of the affinity of the chelators towards Fe and their relative concentration. Bacterial siderophores such as pyoverdine have a higher affinity towards Fe than PS (Yehuda *et al.* 1996). Therefore, if both siderophore and PS are present at the same concentration, Fe is preferentially bound to siderophores; moreover, siderophores may remove Fe from Fe-PS. In contrast to many bacterial siderophores, rhizoferrin from the fungus *Rhizopus arrhizus* has only a slightly higher affinity towards Fe than PS. Iron exchange from rhizoferrin to PS is thought to be the reason for rhizoferrin to be a good Fe source for barley (Yehuda *et al.* 1996). However, not only the affinity of the chelators towards Fe, but also their relative concentration, is important (Yehuda *et al.* 1996). If present at very high concentrations, even chelators with a low affinity towards Fe may exchange Fe from strong chelators. The diurnal rhythm of PS release by the root tips of grasses results in a very high concentration of PS at certain times of the day (Crowley and Gries 1994). Under these conditions PS may be very efficient Fe chelators that could even remove Fe from bacterial siderophores. A high rate of PS release is probably also important for their effectiveness because they are rapidly degraded by microorganisms (Von Wiren *et al.* 1995).

It may be concluded that generally microorganisms compete with plants for Fe. However, some fungi such as *Rhizopus arrhizus* have the potential to increase Fe uptake by plants and could be used as biofertilisers.

Siderophores also play a role in the pathogen-suppressive effect of some biocontrol bacteria (Kloepper and Schroth 1981). The disease suppressive effect of the siderophores is thought to be due to sequestration of Fe by the siderophores of the suppressive bacterium into a form unavailable to the pathogen (Höfte *et al.* 1991, Leong 1986). Siderophores are important in suppression of *Fusarium* spp. and *Erwinia carotovora* (Leong 1986). On the other hand, the control of *Gaeumannomyes graminis* var. *tritici* is mainly due to the antibiotic phenazine, while siderophores play only a minor role (Hamdan *et al.* 1991).

5. THE ROLE OF MICROORGANISMS IN MANGANESE UPTAKE BY PLANTS

Millions of hectares of arable land worldwide are deficient in Mn (Welch 1995). Only the reduced form of Mn (Mn^{2+}) is available to plants, while its oxidised form (Mn^{4+}) is unavailable. Reduction can be either biological or chemical, whereas oxidation is biological (Ghiorse 1988). Thus, the availability of Mn in soils strongly depends on the ratio between Mn-oxidising and Mn-reducing microorganisms (Rengel 1997).

Mn oxidation ($Mn^{2+} + \frac{1}{2}O_2 + H_2O \rightarrow (Mn^{4+}) O_2 + 2H^+$) provides energy to support growth of chemolithotrophic bacteria (Ghiorse 1988). Mn reduction (($Mn^{4+}) O_2 + 4H^+ + 2e^- \rightarrow Mn^{2+} + 2H_2O$) occurs under both anaerobic and aerobic conditions (Trimble and Ehrlich, 1968) and is due to presence of protons and reducing agents carrying electrons (Uren 1981). Mn-reducing microorganisms may use Mn^{4+} in preference to O_2 as terminal electron acceptor (Trimble and Ehrlich 1968).

Crop plant genotypes differ in sensitivity towards Mn deficiency, and rhizosphere microorganisms may play an important role in these genotypic differences (Rengel 1997). Under Mn-deficient conditions, the number of Mn reducers is higher in the rhizosphere of Mn-efficient genotypes than Mn-inefficient wheat genotypes (Rengel *et al.* 1996). Timonin (1946) found greater numbers of Mn-oxidising microbes in the rhizosphere of Mn-inefficient than in Mn-efficient oat cultivars and concluded that this was the basis of genotypic differences in Mn uptake. Later research confirmed these findings and linked greater Mn efficiency to production of root exudates that are toxic to Mn-oxidising microorganisms in the rhizosphere (Timonin 1965).

These differences in Mn efficiency could be exploited in sustainable agriculture by selecting Mn-efficient genotypes in areas with low Mn availability. This would decrease the need for Mn fertilisation and result in more reliable yields.

6.   PLANT GROWTH PROMOTING RHIZOSPHERE ORGANISMS (PGPRs)

Plant growth promoting rhizosphere organisms (PGPRs), are a heterogenous group of microorganisms that stimulate plant growth. They have been isolated from a range of host plants and do not appear to be highly plant species-specific. Several mechanisms of plant growth promotion have been proposed, and a given PGPR often exhibits several positive traits (Höflich *et al.* 1992, de Weger *et al.* 1995, Shishido and Chanway 1999) (Table 2).

Table 2 Mechanisms of plant growth promotion by microorganisms.

Plant growth promotion by microorganisms may be due to...
P solubilisation
N_2 fixation
plant hormone production
pathogen suppression (antibiotics, siderophores),
stimulation of other beneficial microorganisms such as N_2 fixers or mycorrhizal fungi

Phosphorus-solubilising bacteria and associative N_2-fixing bacteria have been particularly well investigated for their potential role in plant nutrition (Crowley and Rengel 1999). In both cases, it appears that the ability of these bacteria to enhance plant nutrition may be confounded in part by growth-promoting effects related to the release of plant hormones or other unidentified growth factors. For example, in a recent study with canola (*Brassica napus*), seven strains of bacteria, selected as the best P-solubilising bacteria among 120 rhizosphere isolates, were tested for their effects on plant growth and P uptake (De Freitas *et al.* 1997). Several of the bacteria increased both plant growth and pod weight, but had no significant effect on P uptake from rock phosphate.

Other data support a more direct role in the contribution of these microorganisms to plant nutrition. Mixed culture inoculations of barley with the associative N_2 fixer *Azospirillum lipoferum* and the P-solubiliser *Agrobacterium radiobacter* have been shown to significantly increase grain yield and N nutrition of plants as compared to inoculation with single cultures (Belimov *et al.* 1995). This appeared to involve a synergistic relationship in which both N_2-fixing activity and rates of P solubilisation were increased by the co-inoculation. In another example of the synergistic relationships that may arise in microbial communities, it was shown that *Bacillus polymyxa* increased *Rhizobium etli* populations and nodulation in the rhizosphere of bean (*Phaseolus vulgaris*) (Petersen *et al.* 1996).

The success of an introduced PGPR is based on a successful colonisation of the rhizosphere. For such colonisation, certain traits are important (de Weger *et al.* 1995, Chin-A-Woeng *et al.* 2000):

- motility and growth rate to match root growth,
- attachment to the root surface (e.g. by polysaccharides),
- capacity to utilise root exudates,
- amino acid synthesis, and
- high competitive ability towards other microorganisms

7. CONCLUSIONS

The rhizosphere is characterised by intense interactions between microorganisms and roots. It plays an important role in soil biological fertility. It is a dynamic system where plant roots can stimulate certain favourable microorganisms, thus creating their specific rhizosphere microflora. Microorganisms are attracted by easily available C sources and signalling substances and may have positive effects on nutrient uptake by the plant by increasing the availability of nutrients. However, they may also compete for nutrients with the plant or reduce nutrient uptake by the roots by degradation or utilisation of solubilising substances.

Even though our knowledge of rhizosphere microbiology is relatively limited, it is clear that rhizosphere microorganisms play a key role in plant productivity and sustainable plant production and should receive far more attention than in the past.

Conventional agriculture uses plant genotypes that produce high yields in combination with a high input of fertilisers and pesticides. The microorganisms in the rhizosphere play only a minor role in crop production, unless they are pathogens. This has led to a selection of plant genotypes which are not well adapted to adverse conditions or low inputs of fertilisers and pesticides (Figure 4).

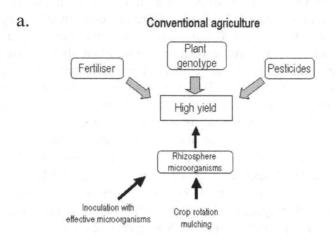

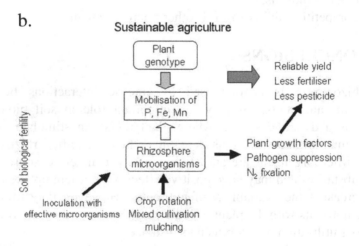

Figure 4 Plant production in (a) conventional and (b) sustainable agriculture.

A sustainable agricultural approach, on the other hand, would recognise the rhizosphere microorganisms as an important factor in crop production (Figure 4). Plant genotypes could be selected for their ability to mobilise nutrients either directly or by stimulating certain rhizosphere microorganisms. This could be further enhanced by management practices such as crop rotation, mixed cultures and mulching, or by inoculating the plants with nutrient mobilising microorganisms. In addition to nutrient mobilisation, rhizosphere microorganisms may increase plant growth by production of plant growth factors, pathogen suppression or N_2 fixation. By exploiting soil biological fertility, yields will be more reliable while using less fertiliser and pesticides.

8.   REFERENCES

Arshad M and Frankenberger W T 1993 Microbial production of plant growth regulators. *In:* Soil Microbial Ecology: Applications in Agricultural and Environmental Management. F B Metting (ed.) pp. 307-348. Marcel Dekker Inc. New York, USA:

Aulakh M S, Wassmann R, Bueno C, Kreuzwieser J and Rennenberg H 2001 Characterization of root exudates at different growth stages of ten rice (*Oryza sativa* L.) cultivars. Plant Biology 3: 139-148.

Banik S and Dey B K 1983 Alluvial soil microorganisms capable of utilizing insoluble aluminium phosphate as a sole source of phosphorus. Zentralblatt fur Mikrobiologie. 138: 437-442.

Bar-Ness E, Hadar Y, Chen Y, Römheld V and Marschner H 1992 Short-term effects of rhizosphere microorganisms on Fe uptake from microbial siderophores by maize and oat. Plant Physiology 100: 451-456.

Belimov A A, Kojemiakov A P and Chuvarliyeva C V 1995 Interaction between barley and mixed cultures of nitrogen fixing and phosphate-solubilizing bacteria. Plant and Soil 173: 29-37.

Bowen G D and Rovira A D 1992 The rhizosphere: the hidden half of the hidden half. *In:* Roots: The hidden half. Y Waisel, A Eshel and U Kafkafi (eds.) pp. 641-669. Marcel Dekker Inc. New York, USA.

Brown M E, Hornby D and Pearson V 1973 Microbial populations and nitrogen in soil growing consecutive cereal crops infected with take-all. Journal of Soil Science 24: 296-310.

Caradus J R 1995 Genetic control of phosphorus uptake and phosphorus status in plants. *In:* Genetic Manipulation of Crop Plants to Enhance Integrated Nutrient Management in Cropping Systems. I. Phosphorus. Proceedings of an FAO/ICRISAT Expert Consultancy Workshop, 15-18 March 1994. C Johansen, K K Lee, K K Sharma, G V Subbarao and E A Kueneman (eds.) pp. 55-74. International Crops Research Institute for the Semi-Arid Tropics. Patancheru Andhra Pradesh, India.

Chin-A-Woeng T F C, Bloemberg G V, Mulders I H M, Dekkers L C and Lugtenberg B J J 2000 Root colonization by phenazine-1-carboxamide-producing bacterium *Pseudomonas chlororaphis* PCL1391 is essential for biocontrol of tomato foot and root rot. Molecular Plant-Microbe Interactions 13: 1340-1345.

Crowley D E and Gries D 1994 Modelling of iron availability in the plant rhizosphere. *In:* Biochemistry of Metal Micronutrients in the Rhizosphere. J A Manthey, D E Crowley and D G Luster (eds.) pp. 199-224. Lewis Publishers. Boca Raton, USA.

Crowley D E, Römheld V, Marschner H and Szaniszlo P J 1992 Root-microbial effects on plant iron uptake from siderophores and phytosiderophores. Plant and Soil 142: 1-7.

Crowley D E and Rengel Z 1999 Biology and chemistry of rhizosphere influencing nutrient availability. *In:* Mineral Nutrition of Crops: Fundamental mechanisms and implications. Z Rengel (ed.) pp. 1-40. The Haworth Press. New York, USA.

Curl E A and Truelove B 1986 The rhizosphere. Springer Verlag. New York.

de Freitas J R, Banerjee M R and Germida J J 1997 Phosphate-solubilizing rhizobacteria enhance the growth and yield but not phosphorus uptake of canola (*Brassica napus* L.). Biology and Fertility of Soils 24: 358-364.

de Weger L A, Van der Bij A J, Dekkers L C, Simons M, Wijffelman C A and Lugtenberg B J J 1995 Colonisation of the rhizosphere of crop plants by plant benficial pseudomonads. FEMS Microbiology Ecology 17: 221-228.

Delhaize E 1995 Genetic control and manipulation of root exudates. *In:* Genetic Manipulation of Crop Plants to Enhance Integrated Nutrient Management in Cropping Systems. I. Phosphorus. Proceedings of an FAO/ICRISAT Expert Consultancy Workshop, 15-18 March 1994. C Johansen, K K Lee, K K Sharma, G V Subbarao and E A Kueneman (eds.) pp. 145-152. International Crops Research Institute for the Semi-Arid Tropics. Patancheru, Andhra Pradesh, India.

Dinkelaker B and Marschner H 1992 *In vivo* demonstration of acid phosphatase activity in the rhizosphere of soil-grown plants. Plant and Soil 144: 199-205.

Duncan R R and Baligar V C 1990 Genetics, breeding, and physiological mechanisms of nutrient uptake and use efficiency: an overview. *In:* Crops as Enhancers of Nutrient Use. V C Baligar and R R Duncan (eds.) pp. 3-36. Academic Press. San Diego, California, USA:

Föhse D and Jungk A 1983 Influence of phosphate and nitrate supply on root hair formation of rape, spinach and tomato plants. Plant and Soil 74: 359-368.

Foster R C 1986 The ultrastructure of the rhizoplane and rhizosphere. Annual Review of Phytopathology 24: 211-234.

Gerke J, Beissner L and Römer W 2000 The quantitative effect of chemical phosphate mobilisation by carboxylate anions on P uptake by a single root. I. The basic concept and determination of soil parameters. Journal of Plant Nutrition and Soil Science 163: 201-212.

Gerke J 1994 Kinetics of soil phosphate desorption as affected by citric acid. Zeitschrift fur Pflanzenernahrung und Bodenkunde 157: 17-22.

Gerretsen F C 1946 The influence of microorganisms on the phosphate intake by the plant. Plant and Soil 1: 51-81.

Ghiorse W C 1988 The biology of manganese transforming microorganisms in soils. In Manganese in Soils and Plants. R D Graham, R J Hannam and N C Uren (eds.) pp. 75-85. Kluwer Academic Publishers. Dordrecht, The Netherlands.

Grayston S J, Wang S, Campbell C D and Edwards A C 1998 Selective influence of plant species on microbial diversity in the rhizosphere.   Soil Biology and Biochemistry 30: 369-378.

Haller T and Stolp H 1985 Quantitative estimation of root exudation of maize plants. Plant and Soil 86: 201-216.

Hamdan H, Weller 'D M and Thomashow L S 1991 Relative importance of fluorescent siderophores and other factors in biological control of *Gaeumannomyces graminis* var. *tritici* by *Pseudomonas flourescens* 2-79 and M4-80R. Applied Environmental Microbiology 57: 3270-3277.

Hendriks L, Claassen N and Jungk A 1981 Phosphatverarmung des wurzelnahen Bodens und P-Aufnahme von Mais und Raps. Zeitschrift fur Pflanzenernahrung und Bodenkunde 144: 486-499.

Hoffland E, Findenegg G R and Nelemans J A 1989 Solubilization of rock phosphate by rape. II. Local root exudation of organic acids as a response to P starvation. Plant and Soil 113, 161-165.

Höflich G, Glante F, Liste H H, Weise I, Ruppel S and Scholz-Seidel C 1992 Phytoeffective combination effects of symbiotic and associative microorganisms on legumes. Symbiosis 14: 427-438.

Höfte M, Seong K Y, Jurkevitch E and Verstraete W 1991 Pyoverdin production by the plant growth-benficial *Pseudomonas* strain 7NSK2: ecological significance in soil. *In* Iron Nutrition and Interactions in Plants. Y Chen and Y Hadar (eds.) pp. 289-297. Kluwer Academic Publishers. Dordrecht, The Netherlands.

Howie W J and Echandi E 1983 Rhizobacteria: Influence of cultivar and soil type in plant growth and yield of potato. Soil Biology and Biochemistry 15: 127-132.

Jiang H-Y and Sato K 1992 Fluctuations in bacterial populations on the root surface of wheat (*Triticum aestivum* L.) grown under different soil conditions. Biology and Fertility of Soils 14: 246-252.

Jiang H-Y and Sato K 1994 Interrelationships between bacterial populations on the root surface of wheat and growth of plant. Soil Science and Plant Nutrition 40: 683-689.

Jones D L and Darrah P R 1994 Amino acid influx at the soil-root interface of *Zea mays* L. and its implications in the rhizosphere. Plant and Soil 163: 1-12.

Jones D L and Edwards A C 1998 Influence of sorption on the biological utilisation of two simple carbon substrates. Soil Biology and Biochemistry 30: 1895-1902.

Jones D L 1999 Amino acid biodegradation and its potential effects on organic nitrogen capture by plants. Soil Biology and Biochemistry 31: 613-622.

Jurkevitch E, Hadar Y and Chen Y 1988 Involvement of bacterial siderophores in the remedy of lime-induced chlorosis in peanut. Soil Science Society of America Journal 52: 1032-1037.

Jurkevitch E, Hadar Y, Chen Y, Chino M and Mori S 1993 Indirect utilisation of the phytosiderophore mugineic acid as an iron source to rhizosphere fluorescent *Pseudomonas*. Biometals 6: 119-123.

Kloepper J W and Schroth M N 1981 Relationship of in vitro antibiosis of plant growth-promoting rhizobacteria to plant growth and displacement of root microflora. Phytopathology 71: 1020-1024.

Kamal K, Hagagg L, Awad L 2000 Improved Fe and Zn acquisition by guava seedlings grown in calcareous soils intercropped with graminaceous species. Journal of Plant Nutrition 23: 2071-2080.

Kumar V and Narula N 1999 Solubilization of inorganic phosphates and growth emergence of wheat as affected by *Azotobacter chroococcum* mutants. Biology and Fertility of Soils 28: 301-305.

Kundu B S and Gaur A C 1980 Effect of phosphobacteria on the yield and phosphate uptake of potato crop. Current Science 49: 159.

Lemanceau P, Corberand T, Gardan L, Latour X, Laguerre G, Boeufgras J M and Alabouvette C 1995 Effect of two plant species, flax (*Linum usitatissimum* L) and tomato (*Lycopersicon esculentum* Mill), on the diversity of soilborne populations of fluorescent pseudomonads. Applied Environmental Microbiology 61: 1004-1012.

Leong J 1986 Siderophores: their biochemistry and possible role in the biocontrol of plant pathogens. Annual Review of Phytopathology 2: 187-209.

Li M, Osaki M, Rao I M and Tadano T 1997-a Secretion of phytase from the roots of several plant species under phosphorus-deficient conditions. Plant and Soil 195: 161-169.

Li M, Shinano T and Tadano T 1997-b Distribution of exudates of lupin roots in the rhizosphere under phosphorus deficient conditions. Soil Science and Plant Nutrition 43: 237-245.

Lynch J M and Whipps J M 1990 Substrate flow in the rhizosphere. Plant and Soil 129: 1-10.

Marilley L and Aragno M 1999 Phytogenetic diversity of bacterial communities differing in degree of proximity of *Lolium perenne* and *Trifolium repens* roots. Applied Soil Ecology 13: 127-136.

Marschner H 1995 Mineral Nutrition of Higher Plants. Academic Press. London, UK.

Marschner P and Crowley D E 1998 Phytosiderophore decrease iron stress and pyoverdine production of *Pseudomonas fluorescens* Pf-5 (pvd-inaZ). Soil Biology and Biochemistry 30: 1275-1280.

Marschner P, Yang C H, Lieberei R and Crowley D E 2001 Soil and plant specific effects on bacterial community composition in the rhizosphere. Soil Biology and Biochemistry 33: 1437-1445.

Martin J K 1971 Influence of plant species and plant age on the rhizosphere microflora. Australian Journal of Biological Sciences 24: 1143-1150.

Meharg A A and Kilham K 1995 Loss of exudates from the roots of perennial ryegrass inoculated with a range of microorganisms. Plant and Soil 170: 345-349.

Merbach W and Ruppel S 1992 Influence of microbial colonization on $^{14}CO_2$ assimilation and amounts of root-borne ^{14}C compounds in soil. Photosynthesis 26: 551-554.

Merbach W, Mirus E, Knof G, Remus R, Ruppel S, Russow R, Gransee A and Schulze J 1999 Release of carbon and nitrogen compounds by plant roots and their possible ecological importance. Journal of Plant Nutrition and Soil Science 162: 373-383.

Mori S 1994 Mechanisms of iron acquisition by graminaceous (strategy II) plants. *In:* Biochemistry of Metal Micronutrients in the Rhizosphere. J A Manthey, D E Crowley and D G Luster (eds.) pp. 225-249. Lewis Publishers. Boca Raton, USA.

Neilands J B 1984 Siderophores of bacteria and fungi. Microbiological Science 1: 9-14.

Neumann G, Massoneau A, Martinoida E and Römheld V 1999 Physiological adaptions to phosphorus deficiency during proteoid root development in white lupin. Planta 208: 373-382.

Petersen D J, Srinivasan M and Chanway C P 1996 *Bacillus polymyxa* stimulates increased *Rhizobium etli* populations and nodulation when co-resident in the rhizosphere of *Phaseolus vulgaris*. FEMS Microbiology Letters 142: 271-276.

Raaijmakers J M, Van der Sluis I, Bakker P A H M, Weisbeek P J and Schippers B 1995 Utilization of heterologous siderophores and rhizosphere competence of fluorescent *Pseudomonas* spp. Canadian Journal of Microbiology 41: 126-135.

Rengel Z 1997 Root exudation and microflora populations in rhizosphere of crop genotypes differing in tolerance to micronutrient deficiency. Plant and Soil 196: 255-260.

Rengel Z, Gutteridge R, Hirsch P and Hornby D 1996 Plant genotype, micronutrient fertilisation and take-all infection influence bacterial populations in the rhizosphere of wheat. Plant and Soil 183: 269-277.

Richardson A E 2001 Prospects for using soil microorganisms to improve the acquisition of phosphorus by plants. Australian Journal of Plant Physiology 28: 897-906.

Richardson A E, Hadobas P A and Hayes J E 2000 Acid phosphomonoesterase and phytase activities of wheat (*Triticum aestivum* L.) roots and utilization of organic phosphorus substrates by seedlings grown in sterile culture. Plant Cell Environment 23: 397-405.

Römheld V 1986 pH changes in the rhizosphere of various crop plants in relation to the supply of plant nutrients. Potash Review 12: 1-12.

Römheld V 1991 The role of phytosiderophores in acquisition of iron and other micronutrients in graminaceous species: an ecological approach. Plant and Soil 130: 127-134.

Römheld V and Marschner H 1990 Genotypical differences among graminaceous species in release of phytosiderophores and uptake of iron phytosiderophores. Plant and Soil 123: 147-153.

Rouatt J W and Katznelson H 1961 A study of the bacteria on the root surface and in the rhizosphere soil of crop plants. Journal of Applied Bacteriology 24: 164-171.

Schachtman D P, Reid R J and Ayling S M 1998 Phosphorus uptake by plants: from soil to cell. Plant Physiology 116: 47-453.

Schaffert R E 1994 Discipline interactions in the quest to adapt plants to soil stresses through nutrient improvement. *In:* Proceedings of the Workshop on Adaptation of Plants to Soil Stresses. INTSORMIL Pub. No. 94-2. pp. 1-16.

Schönwitz R and Ziegler H 1989 Interaction of maize roots and rhizosphere microorganisms. Zeitschrift fur Pflanzenernahrung und Bodenkunde 152: 217-222.

Sheng J and Citovsky V 1996 *Agrobacterium* – plant cell DNA transport: have virulence proteins, will travel. Plant Cell 8: 1699-1710.

Shishido M and Chanway C P 1999 Spruce growth response specificity after treatment with plant growth-promoting *Pseudomonas*. Canadian Journal of Botany 77: 22-31.

Ström L 1997 Root exudation of organic acids: importance to nutrient availability and the calcifuge and calcicole behaviour of plants. Oikos 80: 459-466.

Subbarao G V, Ae N and Otani T 1997 Genotypic variation in iron- and aluminium-phosphate solubilizing activity of pigeonpea root exudates under P deficient conditions. Soil Science and Plant Nutrition 43: 295-305.

Takagi S, Nomoto K and Takemoto T 1984 Physiological aspects of mugineic acid: a possible phytosiderophore of graminaceous plants. Journal of Plant Nutrition 7: 469-477.

Tarafdar J C and Jungk A 1987 Phosphatase activity in the rhizosphere and its relation to the depletion of soil organic phosphorus. Biology and Fertility of Soils 3: 199-204.

Timonin M I 1946 Microflora of the rhizosphere in relation to the manganese - deficiency disease of oats. Soil Science Society of America Proceedings 11: 284-292.

Timonin M I 1965 Interaction of higher plants and soil microorganisms. *In:* Microbiology and Soil Fertility. C M Gilmore and O N Allen (eds.) pp. 135-138. Oregon State University Press. Corvallis, USA.

Toro M, Azcon R and Barea J M 1997 Improvement of arbuscular mycorrhiza development by inoculation of soil with phosphate-solubilizing rhizobacteria to improve rock phosphate bioavailability and nutrient cycling. Applied Environmental Microbiology 63: 4408-4412.

Trimble R B and Ehrlich H L 1968 Bacteriology of manganese nodules. III. Reduction of MnO_2 by two strains of nodule bacteria. Applied Microbiology 16: 695-702.

Trolove S N, Hedley M J, Caradus J R and Mackay A D 1996 Uptake of phosphorus from different sources by *Lotus pedunculatus* and three genotypes of *Trifolium repens*. II. Forms of phosphate utilised and acidification of the rhizosphere. Australian Journal of Soil Research 34: 1027-1040.

Tyler G and Ström L 1995 Differing organic acid exudation pattern explains calcifuge and acidifuge behaviour of plants. Annuals of Botany 75: 75-78.

Uren N C and Reisenauer H M 1988 The role of root exudates in nutrient acquisition. Advances in Plant Nutrition 3: 79-114.

Uren N C 1981 Chemical reduction of an insoluble higher oxide of manganese by plant roots. Journal of Plant Nutrition 4: 65-71.

Vancura V and Hovadik A 1965 Root exudates of plants. II. Composition of root exudates of some vegetables. Plant and Soil 22: 21-32.

Von Wiren N, Mori S, Marschner H and Römheld V 1994 Iron inefficiency in maize mutant ys1 (*Zea mays* L. cv Yellow-stripe) is caused by a defect in uptake of iron phytosiderophores. Plant Physiology 106: 71-77.

Von Wiren N, Römheld V, Shiori T and Marschner H 1995 Competition between micro-organisms and roots of barley and sorghum for iron accumulated in the root apoplasm. New Phytologist 130: 511-521.

Walter A, Römheld V, Marschner H and Crowley D E 1994 Iron nutrition of cucumber and maize: effect of *Pseudomonas putida* YC3 and its siderophore. Soil Biology and Biochemistry 26: 1023-1031.

Wang Y, Brown H N, Crowley D E and Szaniszlo P J 1993 Evidence of direct utilization of a siderophore, ferrioxamine B, in axenically grown cucumber. Plant Cell Environment 16: 579-585.

Warembourg F R and Billes G 1973 Estimating carbon transfers in the plant rhizosphere. *In:* The Soil-Root Interface. J L Harley and RS Scott-Russell (eds.) pp. 183-196. Academic Press. London.

Welch R M 1995 Micronutrient nutrition of plants. Critical Reviews in Plant Science 14: 49-82.

Whipps J M and Lynch J M 1983 Substrate flow and utilization in the rhizosphere of cereals. New Phytologist 95: 605-623.

Whitelaw M A, Harden T J and Helyar K R 1999 Phosphate solubilisation in solution culture by the soil fungus *Penicillium radicum*. Soil Biology and Biochemistry 31: 655-665.

Wiehe W and Höflich G 1995 Survival of plant growth-promoting rhizosphere bacteria in the rhizosphere of different crops and migration to non-inoculated plants under field conditions in north-east Germany. Microbiological Research 150: 201-206.

Yan X, Lynch J P and Beebe S E 1996 Utilization of phosphorus substrates by contrasting common bean genotypes. Crop Science 36: 936-941.

Yehuda Z, Shenker M, Römheld V, Marschner H, Hadar Y and Chen Y 1996 The role of ligand exchange in the uptake of iron from microbial siderophores by gramineous plants. Plant Physiology 112: 1273-1280.

Zhang F S, Ma J and Cao Y P 1997 Phosphorus deficiency enhances root exudation of low-molecular weight organic acids and utilization of sparingly soluble inorganic phosphates by radish (*Raphanus sativus* L.) and rape (*Brassica napus* L.) plants. Plant and Soil 196: 261-264.

Zhu Y, Pierson L S and Hawes M C 1997 Induction of microbial genes for pathogenesis and symbiosis by chemicals from root border cells. Plant Physiology 115: 1691-1698.

Chapter 6
Contributions of Rhizobia to Soil Nitrogen Fertility

Alison McInnes[1, 2] and Krystina Haq[1]

[1] School of Earth and Geographical Sciences, Faculty of Natural and Agricultural Sciences, The University of Western Australia, Crawley, 6009, WA, Australia.
[2] School of Environment and Agriculture, College of Science, Technology and Environment, University of Western Sydney-Hawkesbury, Locked Bag 1797, Penrith South DC, 1797, NSW, Australia.

1. INTRODUCTION

Agricultural systems are dependent on the input of nitrogen (N) to offset losses sustained through produce removal and through processes such as leaching, denitrification and the volatilisation of ammonia (Ladha 1995). A major source of N in agricultural systems is provided through the symbiosis between rhizobia (root nodule bacteria) and legumes. Rhizobia enter legume root systems through root hair infection or through cracks in the epidermis and induce cortical cells to divide and form nodules (Kijne 1992). Within the root nodules, rhizobia transform N between its abundant form as atmospheric gas (N_2) which is metabolically unavailable to plants and metabolically available combined N (NH_3). This process is known as biological N fixation.

After export from the nodule, the fixed N can be used directly for growth by the crop or pasture legume. It has been estimated that N fixation by the legume-rhizobia symbiosis contributes at least 70 million metric tons of fixed N per year into terrestrial ecosystems (Brockwell *et al.* 1995). This

99

L.K. Abbott & D.V. Murphy (eds) Soil Biological Fertility - A Key to Sustainable Land Use in Agriculture. 99-128.

accounts for up to 40% of the total N fixed on earth (Paul and Clark 1996). In areas of arable agriculture, N fixation by crop and pasture legumes is believed to be the dominant N input (Peoples and Craswell 1992) and estimates exceeding 50% of N are frequently reported in the literature (Burns and Hardy 1975, Peoples and Craswell 1992, Unkovich *et al.* 1995, Grey 1999, Peoples and Baldock 2001). In Australia, about 1 x 10^6 t N are fixed annually by crop and pasture legumes, which is equivalent to $1 billion in fertiliser N application (Herridge *et al.* 2001).

Fixed N may also be used indirectly by other plant species through legume root exudates and the breakdown of legume residues. The mineralisation of legume residues contributes varying quantities of N to agricultural systems and is dependent on grazing practices, efficiency of carbon use by decomposers, N demand, plant C:N, lignin:N and polyphenol:N ratios and a range of soil factors (Fillery 2001). In pasture based systems, or where the legume is grown as a green manure, these inputs can be substantial (Brockwell *et al.* 1995). However, even when a large proportion of fixed N is removed as a grain crop, N fixation still improves the N economy of soils (Peoples *et al.* 1995). Legumes commonly increase the yield of subsequent crops by the equivalent of 30-80 kg fertiliser N ha^{-1} and provide the additional advantage of a disease break between crops (Peoples *et al.* 1995).

For the full N benefits to be realised, legumes must be nodulated by rhizobial strains that are effective in their N fixing capacity. In countries with mechanised agricultural production, this is usually achieved through inoculation of legume seed with selected rhizobial strains in the form of commercially available preparations such as gamma-irradiated peat (Date 2001). For example, in Australia, commercial rhizobial strains are selected to be:

- highly effective in their N fixing capacity in association with target host legumes
- capable of persisting in soil over the range of environments in which they are likely to be introduced
- either effective on, or incapable of nodulation with, non-target legumes grown in the same agricultural systems

Stringent quality control of Australian commercially produced inoculants by an independent agency (the Australian Legume Inoculant Research Unit, New South Wales Agriculture, Gosford NSW, Australia) ensures high quality peat inoculants that contain at least 1 x 10^9 viable rhizobia g^{-1}. As a result, farmers have the potential to achieve high rates of N fixation through crop and pasture rotations in agricultural systems.

Unfortunately, legumes do not always become nodulated by the applied inoculant strain. A key limitation that prevents nodule occupancy and N fixation by rhizobial inoculant strains is the presence of background populations of variably effective rhizobia. In the following chapter, we review factors that contribute to the size, strain composition, strain dominance and N fixing effectiveness of background rhizobial populations. We also investigate potential management strategies for enhancing N inputs from fixation through manipulation of background populations to favour nodulation by applied inoculant strains. We conclude by speculating on the impacts that conservation farming practices may have on competition between background populations and introduced inoculant strains.

2. LIMITATIONS TO N FIXATION THROUGH THE PRESENCE OF BACKGROUND POPULATIONS OF RHIZOBIA

High rates of N fixation in agricultural legumes typically occur when either: i) there is no background soil population of rhizobia to compete with introduced inoculant strains or ii) the background population contains rhizobia that nodulate effectively with the target legume species. Where background populations are absent, selection of an effective inoculant strain adapted to the range of soil types over which the target legume is to be grown will ensure high rates of N fixation (Howieson and McInnes 2001). Examples of agricultural legumes that are grown where no background populations of rhizobia are present include *Biserrula pelecinus* and *Stylosanthes seabrana* in Australia (Howieson *et al.* 1995, Date *et al.* 1996) and *Lotus pedunculatus* in South Africa (Jansen van Rensburg and Strijdom 1985).

Where effective background populations of rhizobia are present, inoculation of legumes is unlikely to increase N fixation rates, except where populations are less than 10 rhizobia g^{-1} soil (Thies *et al.* 1991). Accordingly, legumes that nodulate effectively with background populations may only benefit from inoculation at sites where they have never been grown before (Bushby 1982, Thies *et al.* 1991, Bottomley 1992) or where they have not been sown for an extended period. Examples of agricultural legumes that nodulate effectively with background rhizobial populations include lupin and serradella (McInnes 2002, K Haq unpublished data) and subterranean clover (Unkovich and Pate 1998) in south-western Australia, *Vicia faba* in French agricultural soils (Amarger 1988), *Cicer arietinum* in Bangladeshi soils (Sattar *et al.* 1995) and *Phaseolus vulgaris* in Andalusian soils (Rodriguez-Navarro *et al.* 2000).

Where inoculants are added to soils harbouring large background populations of variable effectiveness, a low proportion of nodule occupancy by inoculant strains is the major barrier to increased N input by fixation. For example, Thies *et al.* (1991) showed that 66% nodule occupancy by the inoculant strain was required to significantly increase yield of inoculated compared with uninoculated crops in the presence of background rhizobial populations in Hawaii. In the USA, soybean production areas, attempts to displace dominant and poorly effective *Bradyrhizobium japonicum* serogroup 123 strains by more effective inoculant strains have been unsuccessful (Weber *et al.* 1989, Brockwell *et al.* 1995).

The impracticality of attempting to improve N fixation rates through the introduction of inoculant strains where background populations are present is well recognised (Brockwell *et al.* 1995). In the year of sowing, a high proportion of nodules may be occupied by the inoculant strain because of relatively low background population numbers and locally high numbers of the inoculant strain on seed at the point of root emergence (Brockwell *et al.* 1995). However, in the years following introduction, a pattern of declining nodule occupancy by the introduced strain has been frequently observed (Gibson *et al.* 1976, Roughley *et al.* 1976, Jansen van Rensburg and Strijdom 1985). This can be due to: i) better persistence of locally adapted background strains in the soil environment (Chatel *et al.* 1968), ii) build-up of background strains in the presence of the host rhizosphere (Dowling and Broughton 1986, Bushby 1993, Mendes and Bottomley 1998), iii) greater competitive ability of background strains for nodulation (Dowling and Broughton 1986) and iv) release of large numbers of background strains from senescent nodules (Bushby 1984, Brockwell *et al.* 1987, Thies *et al.* 1995).

The simplest option for improving N fixation in the presence of large populations of variably effective rhizobia is to select an alternative legume species that either nodulates effectively with, or has no interaction with, the background population (Howieson and McInnes 2001). In some cases it may be possible to select a competitive and well-adapted strain from background populations that is capable of dominating nodules in a mixed population. For example, Amarger (1988) found that some effective naturalised strains of rhizobia formed the majority of nodules when inoculated into soils with established background populations between 7×10^3 and 8×10^4 cells g^{-1} soil.

In current research, naturalised eastern Australian populations of rhizobia associated with chickpea, faba bean and medic species are being screened to identify effective strains that dominate nodule populations (McInnes and Thies unpublished data). In Western Australia, several effective and dominant bradyrhizobia have been identified within background populations that have potential as inoculant strains for serradella on acid sands (McInnes 2002). Given that competition for nodulation is one of the major determinants of strain composition in rhizobial populations (Povorov and Vorob'ev 1998 - Section 3.2 - this chapter), selection of effective naturalised strains that are competitive for nodulation under local conditions may represent the best chance for successful inoculant introduction into variably effective background populations.

An alternative strategy to finding locally competitive strains may be the manipulation of background populations to favour greater representation in nodules by introduced and/or effective strain type(s). Successful development of such management strategies will depend on a thorough understanding of the processes that determine the strain composition, strain dominance, genetic diversity and effectiveness of background populations, including:

- how rhizobial populations evolve
- impacts of genetic mutation and DNA exchange on population diversity and effectiveness
- migration of strains between populations
- effects of environmental factors on population size and diversity, the proportion of each strain within populations and the competitive ability of strains for nodulation with the host legume.

3. SOURCES OF GENETIC DIVERSITY WITHIN BACKGROUND RHIZOBIAL POPULATIONS

3.1 Evolution and Genetic Change in Rhizobial Populations

The early evolution of rhizobia probably predates the evolution of terrestrial plants, and the capacity for N fixation is thought to have been acquired during this period (Hennecke *et al.* 1985, Young 1993). The capacity for nodulation evolved subsequently with the appearance of legumes (Sprent 2001) and there is evidence that nodulation genes have spread between different phylogenetic groups of rhizobia by horizontal transfer (Young 1993, Martinez-Romero 1994). Sequencing of *nod* and *nif* genes has revealed that *nod* gene phylogeny is well correlated with the systematics of the host legume, while *nif* gene phylogeny correlates with bacterial phylogeny (Laguerre *et al.* 1996, Povorov 1998).

At the level of local populations of rhizobia infecting a given host legume, a range of mechanisms exist to generate genetic diversity within soil populations, arising from interactions between the DNA of different organisms and environmental variables (Figure 1). Rhizobial genomes have been shown to change *in vitro* through mutation, genomic rearrangement and the acquisition of DNA by the normal bacterial processes of transformation, transduction and conjugation (Dowling and Broughton 1986, Martinez *et al.* 1990, Barnet 1991, Young 1993). DNA rearrangement can occur during cell replication and when various mobile plasmids (Trevors *et al.* 1987) and insertion elements (Ronson 1999) are transmitted between bacteria in close contact with each other. The rates at which these events occur will be mediated by environmental and internal mutagens, rates of cell division and environmental factors that enhance genetic recombination.

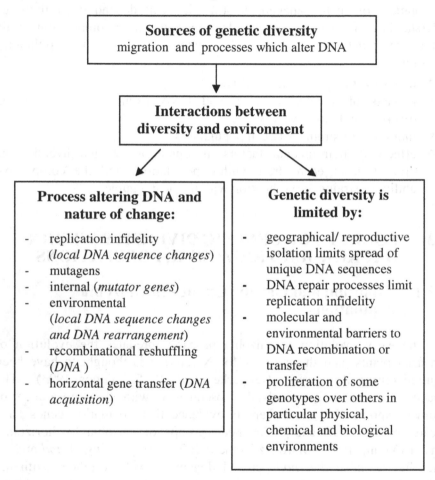

Figure 1 The evolution of genetically diverse bacterial populations arises from a balance between processes producing diversity and processes which act to limit variation. Interactions between genetic diversity and the environment drive the outcome of these processes (modified from Arber 2000).

In controlled experiments in soil, recombination rates have been shown to be affected by presence of clay minerals, presence of solid interfaces, pH, temperature, humidity, oxygen availability, nutrient status and availability of energy sources (Klingmuller *et al.* 1990, van Elsas and Trevors 1990). Field studies have shown that transfer of DNA between organisms is much less likely to occur in soil than in controlled laboratory studies (Hirsch and Spokes 1994). This may be partly due to the isolation of different genotypes at different microsites in the soil environment, and to molecular and genetic barriers to gene transfer that limit this process. Barriers include mechanisms that prevent the stable incorporation of transferred DNA and the expression of characters conferring traits with selective advantage (Figure 1).

Plasmid exchange has been studied extensively because of the importance of *Sym* plasmids, which carry the *nod* and *nif* genes for many species of *Rhizobium* (Martinez *et al.* 1990). Plasmid exchange may be a more important source of genetic variation in *Rhizobium* than in *Bradyrhizobium*, because, although plasmid transfer occurs between bradyrhizobia in soil (Kinkle *et al.* 1993), plasmids in bradyrhizobia can be scarce (Martinez-Romero 1994).

Sullivan and Ronson (1998) showed that chromosomally-located symbiotic genes in a *Mesorhizobium loti* inoculant strain could be transferred into non-symbiotic rhizobia (Sullivan *et al.* 1996). They proposed that a genetically diverse population recovered from the nodules of *Lotus corniculatus* inoculated seven years previously in a soil devoid of background *M. loti* (Sullivan *et al.* 1995) arose via this mechanism. This mechanism may also be important for the development of genetic diversity within bradyrhizobial populations (bradyrhizobia have chromosomally-located symbiotic genes), although the presence of non-symbiotic populations of *Bradyrhizobium* is yet to be reported. However, several studies have identified oligotrophic soil bacteria that are phylogenetically related to *Bradyrhizobium* (on the basis of 16S rRNA sequences) and have the potential to act as the recipients of transferred symbiotic genes (Young and Haukka 1996, Saito *et al.* 1998). Genetically diverse populations of highly effective bradyrhizobia associated with inoculated serradella and lupin in Western Australian soils (McInnes 2002, Haq unpublished data) may have arisen via this mechanism.

The development of genetic variation in rhizobia can be rapid, as found from studies of rhizobial inoculant strains recovered from the field after 1-20 years (Diatloff 1977, Brunel *et al.* 1988, Lochner *et al.* 1989, Lindstrom *et al.* 1990, Gibson *et al.* 1991, Kay *et al.* 1994, Sanginga *et al.* 1994, Santos *et al.* 1999). The majority of field reisolates, characterised by multiple strain typing methods, were identical to the original inoculant strains. However, evidence for genetic change in inoculant strains was found in all studies but one (Sanginga *et al.* 1994). Field isolates of inoculant strains varied in theirphenotypic characteristics and competitive ability, with minor and major genotypic variation apparent using RFLP (Lindstrom *et al.* 1990), RAPD PCR (Kay *et al.* 1994) and REP and ERIC PCR (Santos *et al.* 1999) fingerprinting.

Some field isolates showed reduced or increased capacity to fix N relative to laboratory cultures of the parent inoculant strain (Diatloff 1977, Gibson *et al.* 1991, Santos *et al.* 1999). These studies indicate that all introduced inoculant strains are likely to undergo genotypic change after release into the soil and that this will occur within the time frame of many research programs.

Studies of rhizobial populations in nodules have concluded that little recombination occurs between organisms within nodule tissue (Young 1985, Young and Wexler 1988, Harrison *et al.* 1989). However, studies of soil populations have shown that significant recombination occurs in the soil environment (Segovia *et al.* 1991, Gordon *et al.* 1995, Louvrier *et al.* 1996). These apparently contradictory findings indicate that rhizobial populations in nodules comprise only a small proportion of total population diversity at a site and that clonal increase of certain genotypes in nodules contributes significantly to population structure in nodules of a succeeding host legume.

3.2   Mechanisms that Determine Effectiveness in Rhizobial Populations

The effectiveness of strains in background populations may be influenced by the presence of the host legume. Two early Australian studies (Robinson 1969-a, 1969-b) suggested that subterranean clover and red clover selected effective N fixing rhizobia from mixed populations. However, subsequent work has shown that there is usually no correlation between the N fixing capacity of rhizobial genotypes and their potential to be preferentially selected by the host (Hynes and O'Connell 1990, Povorov 1998, Denison 2000). The more recent view is that N fixing effectiveness is indirectly selected for by a range of legume imposed "sanctions" against ineffective strains in nodules (Denison 2000). These sanctions (e.g. reduced oxygen supply to nodules, production of acid hydrolases, the linking of bacterial carbon metabolism with host plant N metabolism; Udvardi and Khan 1992) may result in preferential allocation of resources to effective bacteroids and/or nodules (Jiminez and Casadesus 1989). The net effect is an increase in favoured genotypes in the soil population through release from senescent nodules. Denison (2000) argues that these mechanisms are likely to influence population dynamics of rhizobia infecting legumes with determinate nodules rather than indeterminate nodules. As bacteroids from determinate nodules remain viable following nodule senescence, legume sanctions against ineffective bacteroids in nodules will translate to a smaller proportion of these genotypes being released into the soil population. The situation with indeterminate nodules is more complex. As these bacteroids lose viability following nodule senescence, sanctions against ineffective genotypes do not affect the soil population. However, undifferentiated bacteria from nodules do contribute to the soil population.   Nodule occupancy by more than one genotype is a relatively common event (Thies *et al.* 1992,   Brockwell *et al.*1995) and means that parasitic rhizobia may proliferate in nodules containing effective bacteroids against which no legume sanctions are applied.   In this case, host-mediated selection for effective genotypes in soil is unlikely to occur.

Alternative views suggest that strain composition within nodules is unrelated to their effectiveness in association with the host legume. Adaptations favouring the long-term survival of rhizobia in soil may be more important drivers of natural selection for the strain composition of populations than their capacity to fix N with the host legume (Sprent 1994, Howieson and McInnes 2001). A modelling study by Povorov and Vorob'ev (1998) suggests that the single greatest influence on the microevolution of naturalised rhizobial populations may be competition for nodulation. Either of these mechanisms would partially account for the observed variability in effectiveness associated with many background populations (Section 2 this chapter).

3.3   Migration of Rhizobial Strains between Populations

Diversity within rhizobial populations may also arise due to migration of genetically distinct groups into an area, as evidenced by the widespread distribution of many rhizobial lineages (Martinez-Romero and Caballero-Mellado 1996). Mechanisms for dispersal of rhizobia may include transport on soil in wind and surface water and spread by migratory and domesticated animals and man (Parker *et al.* 1977,   Barnet 1991,   Strain *et al.* 1995). Rhizobia have been shown to survive in soil and dust particles on seed for extended periods (Parker *et al.* 1977, Perez-Ramirez *et al.* 1998) and diverse strains associated with agricultural legume species may have been introduced into soil populations in this way. Consequently, rhizobial populations may consist of strains that have co-evolved with the indigenous legumes at a site and/or strains that have arrived subsequently through the agencies described above.

4.   TRENDS IN STRAIN RICHNESS AND GENETIC DIVERSITY WITHIN RHIZOBIAL POPULATIONS

The processes of evolution, genetic change and migration described above have resulted in the development of rhizobial populations throughout the world that are typically diverse in both the number of different strains

they contain (strain richness) and the genetic variation between strains
(McInnes 2002). Some species harbour inherently strain-rich populations of
rhizobia in their nodules relative to other species (e.g. *Phaseolus coccineus*;
Souza *et al.* 1994) and, for a given species, strain richness in nodules may
vary greatly between sites (Thurman and Bromfield 1988, Strain *et al.*
1994). Populations of nodule isolates that show high strain richness may
(Young 1985, Young *et al.* 1987) or may not (Souza *et al.* 1994) be
genetically diverse, as determined from the analysis of multilocus enzyme
electrophoresis profiles (Selander *et al.* 1986). Genetic variability in nodule
populations, like strain richness, varies between species or cultivars grown at
the same site (Caballero-Mellado and Martinez-Romero 1999) and between
different sites sown to the same species (Silva *et al.* 1999).

Within these diverse populations, a relatively small proportion of strains
are recovered from nodules at high frequency (Turco and Bezdicek 1987,
Thurman and Bromfield 1988, Hartman and Amarger 1991). Typically,
only one or two strains occupy more than 30% of nodules in published
population studies (McInnes 2002). This population trait creates a challenge
for the development of management strategies to enhance representation of
preferred rhizobial strains in soil and/or host nodule populations. However,
it may be possible to manipulate environmental factors that impact on the
size, diversity and strain dominance within rhizobial populations and thus
improve representation by preferred strains. These environmental factors are
reviewed in Table 1.

5.   ENVIRONMENTAL FACTORS KNOWN TO AFFECT RHIZOBIAL POPULATIONS IN SOIL

Rhizobial populations in agricultural soils can be affected by many
factors including the introduction of inoculant strains, the choice of host
legume and non-legume plant species, impacts of soil type and climate and
agricultural management practices (Vincent 1965, Parker *et al.* 1977,
Lowendorf 1980, Bushby 1982, Dowling and Broughton 1986, Barnet
1991, Bottomley 1992, Brockwell *et al.* 1995).

At the whole-population level, environmental factors may affect
population size (Table 1) and overall diversity (e.g. Palmer and Young
2000). For example, an extensive study of rhizobial populations in a range
of diverse environments on the island of Maui, Hawaii, USA (Woomer and
Bohlool 1989) found that 95% of observed variation in the size of
indigenous populations was accounted for by a model including legume
density, mean annual rainfall and total extractable bases. Within rhizobial
populations, the relative persistence of individual strains may be affected by
many environmental factors, and strains may be altered in their competitive

ability for nodulation with the resident host legume (Table 1). Introduced strains are likely to respond to different sets of factors than background rhizobia that are adapted to the local environment.

In the Maui study (Woomer and Bohlool 1989), populations kinetics of introduced rhizobia were shown to be a function of soil organic carbon, soil water holding capacity and cation exchange capacity.

Some of the environmental factors known to affect rhizobial population size and strain composition (e.g. soil type, temperature and moisture content; see Table 1) cannot be readily manipulated by farmers and offer no prospect for enhancing nodule occupancy by introduced inoculant strains. Potentially suitable factors include choice of host and non-host plant species (crop rotations, pasture composition), soil pH (lime application, use of non-acidifying fertilisers), soil nutrient status (fertiliser N, P, K, Ca and Fe application), soil organic content (pasture/green manure phases, stubble retention) and tillage practices (Table 1). Manipulation of the composition of rhizobial populations through the application of herbicides, pesticides and fungicides may also be feasible but has not (to the authors knowledge) been investigated. Manipulation of non-rhizobial components of the soil biota that have either beneficial or deleterious effects on rhizobial populations (e.g. bacteria, fungi, protozoa, phage and *Bdellovibrio*, see Table 1) may represent another tool for managing rhizobial populations (Bowen and Rovira 1999), although their interactions in the soil environment are currently poorly understood.

Enhancement of nodulation by effective strains may also be possible through management strategies that specifically target the infection process, for example:

- dual inoculation of rhizobial strains on seed with beneficial PGPRs that enhance inoculant survival and infection through the production of hormones, siderophores, antibiotics, phosphate solubilising enzymes, compounds that stimulate root hair growth etc. (Bowen and Rovira 1999)
- selection of inoculant strains with superior motility and cell surface characteristics that lead to early nodule initiation (Lupwayi *et al.* 1996)
- selection of bacteriocin-producing inoculant strains (Dowling and Broughton 1986) that may inhibit background strains
- use of legume cultivars that selectively enhance nodulation by the applied inoculant strain (e.g. through isoflavonoid production; Dakora and Phillips 1996)
- genetic modification of hosts and/or inoculant strains to improve the outcome of competition for nodulation and/or N-fixing capacity (Bosworth *et al.* 1994, Phillips and Streit 1998).

Table 1 Environmental factors affecting population size, strain persistence and nodulation.

Environmental factor		Population size and strain persistence	Nodulation (nodule formation, interstrain competition and host effects)
Legume host species			
	Rhizosphere effects	Rhizobial population size can be enhanced in the presence of the homologous host legume (1, 2, 3, 4, 5, 6, 7, 8, 9, 10). Non-host legume species may also enhance population size (11, 12). Within host legume rhizospheres there may be specific enhancement of some strain types (13).	Different host species or cultivars may favour nodule occupancy by selected strain types (8, 14, 15). Some hosts select effective strains from mixed populations (16, 17, 18).
	Nodule senescence	Rhizosphere populations can be increased through the release of rhizobia from senescent nodules (5, 8, 10, 22, 23).	There may be local enrichment of some strain types from nodule senescence in undisturbed soils (8).
Non-host plants		Non-legumes can enhance rhizobial population size in the rhizosphere relative to the bulk soil – but not as much as the homologous host legume (1, 11, 12, 22, 24). Rhizobial numbers in non-host rhizosphere may be reduced due to toxic effects of root exudates, root decomposition products or the associated microflora (12, 25).	Non-legumes can reduce soil N levels and promote nodulation (9).
Soil physics			
	Soil type	Soil type may influence population size (26) and the persistence of individual strain types (27, 28).	Strain competitive ability is affected by different soil types (29, 30).
	Clay content	Clay can provide protection from desiccation and heat for *Rhizobium* spp. (11, 12, 22, 31, 34, 35), from predation (7) or from competition with other rhizobia (36).	Clay content can alter strain competitive ability (8, 22).

Table 1 Continued on Page 111

Table 1 continued.

Environmental factor		Population size and strain persistence	Nodulation (nodule formation, interstrain competition and host effects)
Soil physics continued			
	Soil texture	Population size and strain persistence is higher in heavier textured soils (22, 31). Rhizobia may be protected from predation and low soil water potential inside soil aggregates or small pores (7, 32, 33). Poor survival may be associated with water deficit and nutrient deficiencies in coarse textured soils, aluminium and manganese toxicity (acid soils), high P binding capacity and variable structure during wetting/drying cycles in fine textured soils (8).	Strain competitive ability is observed to vary with soil texture (22).
	Temperature	High temperature affects strain persistence and population size (7, 8, 11, 37, 38, 39) and capacity to survive varies between strains and different taxonomic groups (1, 2, 3, 5, 22, 31). Some studies show that bradyrhizobia can be poorly tolerant of hot and/or dry conditions (28, 40, 41, 42).	Temperature affects nodule occupancy by different strain types (8, 14, 43, 44) and nodule formation (1, 22). Nodule initiation in serradella and lupin may be reduced at low temperature (45, 46).
	Soil moisture	Low soil moisture adversely affects strain persistence, population size and rhizobial movement in soil (5, 7, 8, 12, 22, 37). Reduced numbers occur under waterlogging in some studies (11, 22, 23) but not others (9, 12). Population size has been correlated with mean annual rainfall (38).	There is reduced nodulation under low moisture conditions (1, 8, 9, 22) and strain competitive ability may be altered (5) or unaltered (47).
Soil chemistry			
	Fertility		Nodule occupancy by inoculant strains correlates with soil fertility in some studies (44, 48).

Table 1 Continued on Page 112

Table 1 continued.

Environmental factor	Population size and strain persistence	Nodulation (nodule formation, interstrain competition and host effects)
Soil chemistry continued		
Acidity	Acidity affects colonisation of soil and rhizosphere in acid sensitive strains (12, 22, 49) particularly below pH 4 (7). Low survival in acid soils may be associated with low P, low Ca and high Al, and strain tolerance to these conditions can differ within the same taxonomic group (8).	Strains show different competitive abilities in acid soils (22, 50) and nodule formation may be affected (8).
Alkalinity	Bradyrhizobial growth can be inhibited at high pH (22, 23)	Strain competition can be altered under alkaline conditions (5)
Aluminium	Al may (8) or may not (5, 51) be toxic to rhizobia in acid soils (8)	Al can affect nodule formation (8)
Boron		Bo is required for nodule development (9, 52).
Calcium	There is an interaction of Ca and low pH on the survival of some rhizobia in acid soil in that Ca can ameliorate the effects of low pH and aluminium and manganese toxicity (22, 52). High Ca in alkaline soils may impact on *B*. sp. (*Lupinus*) survival (53).	Ca influences nodulation of lupin and other species (9, 53). Ca interacts with low pH to alter the competitive ability of some strain types in acid soil (8, 22, 52, 54).
Growth (C/N) substrates	Different rhizobial taxa, have variable growth substrate requirements (7, 55).	
Iron	Survival of rhizobia in soil is unlikely to be affected by soil Fe content (54, 56).	Fe availability can alter the capacity for bradyrhizobia to nodulate lupin and other species (56, 57, 58).
Manganese	Toxic levels of Mn in acid soils may affect strain persistence (12).	

Table 1 Continued on Page 113

Table 1 continued.

Environmental factor	Population size and strain persistence	Nodulation (nodule formation, interstrain competition and host effects)
Soil chemistry continued		
Nitrogen	High levels of N affect rhizosphere colonisation (76) but high N soils can still support large populations (23).	Strains vary in their competitive ability in the presence of high N (5, 22) which has the capacity to suppress nodulation (1, 5, 9).
Other nutrients	Free living rhizobia also require S, K, Mg, Cu, Zn, Mo, Ni, Co and Se for growth (60)	
Organic content	High soil organic content improves strain survival and population size (8, 22, 38) possibly by increasing soil moisture content (22).	Strain competitive ability may vary with organic content (22).
Phosphorus	P can influence the growth and survival of rhizobia (9). Population size has been correlated with P retention by soil (38).	P supply affects strain competitive ability (22, 62, 63). P-sequestering strains can dominate nodules in a P deficient soil (8). Nodule formation may be reduced under low P conditions (1, 9, 64).
Potassium	K availability influences composition of soil populations (11).	K supply affects strain competitive ability (22).
Salinity	Strains vary in their salt sensitivity (8, 22, 65) with WA bradyrhizobia from *Acacia* spp. being more salt sensitive than rhizobia in culture (42). However, many rhizobia can grow and survive at salt concentrations which are inhibitory to agricultural legumes (66).	Salinity may inhibit nodulation (5, 22).

Table 1 Continued on Page 114

Table 1 continued.

Environmental factor	Population size and strain persistence	Nodulation (nodule formation, interstrain competition and host effects)
Soil biology		
Other rhizobia	Other rhizobia may be mutually antagonistic through the production of bacteriocins (5, 23). It is likely that rhizobial strains will compete with each other (and other soil microorganisms) for limited growth substrates (7, 11).	Bacteriocin production can affect strain competitive ability (5).
Other bacteria	May be antagonistic or stimulatory to rhizobia (1, 5, 11, 17) but effects have not been conclusively shown directly in soil (7, 12, 67).	Soil bacteria may affect strain competitive ability (8, 68).
Soil fungi/mycorrhiza	Soil fungi inhibit rhizobial growth in culture (12).	Mycorrhizal formation acts synergistically with nodulation to improve N fixation where P is limiting (9). There is no knowledge of the effects of mycorrhizal formation on competition (5).
Protozoa/*Sitona* weevils	Protozoa reduce rhizobial population size (5, 7, 8, 17, 22).	Protozoa inhibit nodulation (17). *Sitona* weevils feed on nodules (9, 69) and may reduce strain enrichment through nodule senescence.
Parasites – *Bdellovibrio*, phage	There is no strong evidence that parasites reduce rhizobial numbers in soil (1, 5, 7, 12). *Bdellovibrio* isolated from WA soils were not parasitic on *B.* sp. (*Lupinus*) (11).	There is no evidence of the effects of parasites on competition and nodulation (5).
Actinomycetes		Actinomycetes can inhibit nodulation (5).
Collembola		Collembola may affect rhizobial competitive ability (70)

Table 1 Continued on Page 115

Table 1 continued.

Environmental factor		Population size and strain persistence	Nodulation (nodule formation, interstrain competition and host effects)
Agricultural practices			
	Pesticides	Application of fungicides and insecticides to seed affects inoculant strain survival (1, 5, 9, 25, 37). Herbicides may decrease the growth of rhizobia (25).	
	Phosphate fertiliser	P fertiliser impacts on rhizobial survival through acidifying effects and through toxicity of associated heavy metals (1).	P fertiliser influences competition for nodulation in combination with low pH conditions (5).
	Seed coating	Sodium molybdate dressings can affect rhizobial survival (9).	
	Pasture/tillage	Higher population numbers have been measured in no-till sites compared with conventionally tilled sites (25, 39) and under pasture or legume and cereal crops compared with fallow (13, 71).	
	Liming	Liming adversely effects *B.* sp. (*Lupinus*) survival on seed and in soil (53, 73, 74) but is beneficial for acid sensitive rhizobia (9, 12, 22, 75). Population numbers may remain unchanged in soil with liming (8) but liming can improve rhizosphere colonisation (9).	Liming improves nodule formation in acid sensitive strains (22) and can cause nodule occupancy by different strain types (5, 8, 77).

Table 1 Continued on Page 116

Table 1 continued.

References

1; Vincent 1965, 2; Chatel and Parker 1973-a, 3; Chatel and Parker 1973-b, 4; Chowdhury et al. 1968, 5; Dowling and Broughton 1986, 6; Woomer et al. 1990, 7; Barnet 1991, 8; Bottomley 1992, 9; Brockwell et al. 1995, 10; Thies et al. 1995, 11; Parker et al. 1977, 12; Lowendorf 1980, 13; Mendes and Bottomley 1998, 14; Trinick 1985, 15; Bottomley et al. 1994, 16; Robinson 1969-b, 17; Jones 1983, 18; Weaver et al. 1989, 19; Mhamdi et al. 1999, 20; Minamisawa et al. 1999, 21; Thurman and Bromfield 1988, 22; Bushby 1982, 23; Hirsch 1996, 24; Leung et al. 1994, 25; Slattery et al. 2001, 26; Issa and Wood 1995, 27; Crozat et al. 1987, 28; Woomer 1990, 29; Johnson and Means 1963, 30; Schmidt 1988, 31; Marshall 1964, 32; Bottomley and Dughri 1989, 33; Rutherford and Juma 1992, 34; Bushby and Marshall 1977, 35; Heijnen et al. 1992, 36; Heijnen et al. 1993, 37; Alexander 1985, 38; Woomer et al. 1988, 39; Hungria and Vargas 2000, 40; Miller and Pepper 1988, 41; Barnet and Catt 1991, 42; Marsudi et al. 1999, 43; Turco and Bezdicek 1987, 44; Thies et al. 1992, 45; J.G. Howieson unpublished data, 46; Peltzer 1996, 47; Leung and Bottomley 1994, 48; Wolff et al. 1991, 49; Howieson and Ewing 1986, 50; Young 1990, 51; Flis et al. 1993, 52; Munns 1977, 53; Howieson et al. 1998, 54; O'Hara et al. 1988, 55; Chakrabarti et al. 1981, 56; Abd-Alla 1998, 57; Tang et al. 1992, 58; O'Hara et al. 1993, 59; Abd-Alla 1999, 60; O'Hara et al. 2001, 61; Caballero-Mellado and Martinez-Romero 1999, 62; Almendras and Bottomley 1987, 63; Almendras and Bottomley 1988, 64; Mullen et al. 1988, 65; Moawad and Beck 1991, 66; Singleton et al. 1982, 67; Trinick et al. 1983, 68; Fuhrmann and Wollum 1989, 69; Gibson 1977, 70; Lussenhop 1993, 71; Dye et al. 1995, 72; Coutinho et al. 1999, 73; Parker and Oakley 1965, 74; Chatel 1969, 75; Slattery and Coventry 1993, 76; Brockwell et al. 1989, 77; Dughri and Bottomley 1984.

The most successful management options are likely to combine strategies that reduce the size and diversity of background populations with strategies that favour nodulation by target inoculant strains. Factors known to reduce the size of rhizobial populations overall, such as absence of the host legume, production of toxic root exudates, low P availability, tillage and fallowing (Table 1), may assist in reducing background populations prior to the introduction of an inoculant strain. Factors known to reduce the genetic diversity of background populations (choice of host, soil acidity, soil Mg content, N or NPK fertiliser application, tillage; Thurman and Bromfield 1988, Young 1990, Caballero-Mellado and Martinez-Romero 1999, Coutinho *et al.* 1999, Mhamdi *et al.* 1999, Minamisawa *et al.* 1999, Hungria and Vargas 2000, Palmer and Young 2000) may enhance the establishment and competitive ability of applied inoculant strains.

Given the complexity of background populations, and the capacity for genetic change even in introduced inoculant strains (Section 3.1 this chapter), intervention to enhance nodulation by desired strains will need to be ongoing. Applied management strategies may be more successful in cropping systems than in permanent or long-term pasture phases as there is more opportunity for intervention. As the rhizobial strains occurring in geographically separated populations are often different (Strain *et al.* 1995, Aguilar *et al.* 1998, Handley *et al.* 1998, Mhamdi *et al.* 1999, McInnes 2002), the success of strategies chosen to improve nodulation and N fixation of legumes grown in soils with background populations will probably be site-dependent.

6. PRACTICES AFFECTING RHIZOBIAL POPULATIONS

From our knowledge of the factors that affect rhizobial population size, diversity and strain dominance reviewed above, we can speculate on the impacts that conservation farming will have on background rhizobial populations and on the successful introduction of commercial inoculant strains. Practices such as reduced tillage and stubble retention lead to increased soil organic matter, improved soil structure, improved aeration and reduced waterlogging and salinity, all of which are likely to promote rhizobial survival and increase the size of background populations (Table 1). Longer legume phases in rotations (pasture and forage legumes) may have a similar effect, through increases in soil organic matter and the presence of a host legume (Table 1).

Increases in the size of background populations are likely to make successful introduction of inoculant strains more difficult to achieve (Section 2 this chapter). Increased strain richness and genetic diversity under a greater range of plant species in cropping rotations may have a similar effect. The effect of reduced tillage on rhizobial populations is likely to be complex. Reduced tillage means that introduced inoculant strains will be unevenly distributed through the soil profile and may fail to colonise the soil environment beyond the point of introduction. However, Slattery *et al.* (2001) suggested that the absence of mechanical cultivation should assist rapid soil colonisation by inoculant strains (by reducing adverse soil conditions resulting from tillage). An increase in background rhizobial populations is likely under minimal or no tillage, and these populations may (Hungria and Vargas 2000) or may not (Palmer and Young 2000) be less genetically diverse than background populations in tilled systems.

Conservation farming practices may also have an impact on the genetic composition of strains in nodules, and on their N fixing effectiveness. A more buffered soil environment is likely to impose less stringent environmental selection for saprophytic survival in rhizobia, and therefore host legume mediated selection may have a greater influence on the strain composition of rhizobial populations. This may or may not favour populations that fix N effectively (Section 3.3 this chapter). The strain composition of rhizobial populations may also shift in response to practices such as reduced application rates of fertiliser and other agrichemicals (e.g. herbicides and fungicides).

7. CONCLUSION

In this chapter, we have reviewed the factors that contribute to the size and diversity of rhizobial populations and have suggested a range of management strategies that could be explored to enhance the establishment of rhizobial inoculant strains in soils containing background populations. Given: i) the complexity of the mechanisms that determine the strain composition, strain dominance, genetic diversity and effectiveness of background populations, ii) the variability in rhizobial populations at different locations, and iii) the capacity for rapid evolution of new strains through genetic change and exchange (including changes in introduced inoculant strains), there are considerable barriers to overcome to improve nodulation by applied inoculant strains.

Successful management strategies are likely to include: i) selection of dominant and effective inoculant strains originating from soil types in which target legumes are to be sown, ii) reduction of background population size and diversity prior to sowing and iii) a selective advantage for nodulation by the applied inoculant strain. Changing farm practices, such as the shift toward conservation tillage, need to be taken into account in the development of new inoculation strategies.

8. REFERENCES

Abd-Alla M H 1998 Growth and siderophore production *in vitro* of *Bradyrhizobium* (Lupin) strains under iron limitation. European Journal of Soil Biology 34: 99-104.

Abd-Alla M H 1999 Nodulation and nitrogen fixation of *Lupinus* species with *Bradyrhizobium* (lupin) strains in iron-deficient soil. Biology and Fertility of Soils 28: 407-415.

Aguilar O M, Grasso D H, Riccillo P M, Lopez M V and Szafer E 1998 Rapid identification of bean *Rhizobium* isolates by a *nif* H gene-PCR assay. Soil Biology and Biochemistry 30: 1655-1661.

Alexander M 1985 Ecological constraints on nitrogen fixation in agricultural ecosystems. *In:* Advances in Microbial Ecology. Volume 8. K C Marshall (eds.) pp. 163-183. Plenum Press. New York.

Almendras A S and Bottomley P J 1987 Influence of lime and phosphate on nodulation of soil grown *Trifolium subterraneum* L. by indigenous *Rhizobium trifolii*. Applied and Environmental Microbiology 53: 2090-2097.

Almendras A S and Bottomley P J 1988 Cation and phosphate influences on the nodulating characteristics of indigenous serogroups of *Rhizobium trifolii* on soil grown *Trifolium subterraneum* L. Soil Biology and Biochemistry 20: 345-351.

Amarger N 1988 The microbial aspects of faba bean culture. *In:* Nitrogen Fixation by Legumes in Mediterranean Agriculture. D P Beck and L A Materon (eds.) pp. 173-178. ICARDA. Syria.

Arber W 2000 Genetic variation: molecular mechanisms and impact on microbial evolution. FEMS Microbiology Reviews 24: 1-7.

Barnet Y M 1991 Ecology of legume root nodule bacteria. *In:* Biology and Biochemistry of Nitrogen Fixation. M J Dilworth and A R Glenn (eds.) pp. 199-228. Elsevier. Amsterdam.

Barnet Y M and Catt P C 1991 Distribution and characteristics of root-nodule bacteria isolated from Australian *Acacia* spp. Plant and Soil 135: 109-120.

Bosworth A H, Williams M K, Albrecht K A, Kwiatkowski R, Beynon J, Hankinson T R, Ronson C W, Cannon F, Wacek T J and Triplett E W 1994 Alfalfa yield response to inoculation with recombinant strains of *Rhizobium meliloti* with an extra copy of *dct*ABD and/or modified *nif*A expression. Applied and Environmental Microbiology 60: 3815-3832.

Bottomley P J 1992 Ecology of *Bradyrhizobium* and *Rhizobium*. *In:* Biological Nitrogen Fixation. G Stacey, R H Burris and H J Evans (eds.) pp. 293-348. Chapman and Hall. New York.

Bottomley P J, Cheng H H and Strain S R 1994 Genetic structure and symbiotic characteristics of a *Bradyrhizobium* population recovered from a pasture soil. Applied and Environmental Microbiology 60: 1754-1761.

Bottomley P J and Dughri M H 1989 Population size and distribution of *Rhizobium leguminosarum* bv. *trifolii* in relation to total soil bacteria and soil depth. Applied and Environmental Microbiology 55: 959-964.

Bowen G D and Rovira A D 1999 The rhizosphere and its management to improve plant growth. Advances in Agronomy 66: 1-102.

Brockwell J, Bottomley, P and Thies J E 1995 Manipulation of rhizobia microflora for improving legume productivity and soil fertility: A critical assessment. Plant and Soil 174: 143-180.

Brockwell J, Gault R R, Morthorpe L J, Peoples M A, Turner G L and Bergersen F L 1989 Effects of soil nitrogen status and rate of inoculation on the establishment of populations of *Bradyrhizobium japonicum* and on the nodulation of soybeans. Australian Journal of Agricultural Science 40: 753-762.

Brockwell J, Roughley R J and Herridge D F 1987 Population dynamics of *Rhizobium japonicum* strains used to inoculate three successive crops of soybeans. Australian Journal of Agricultural Research 38: 61-74.

Brunel B, Cleyet-Marel J C, Normand P and Bardin R 1988 Stability of *Bradyrhizobium japonicum* inoculants after introduction into the soil. Applied and Environmental Microbiology 54: 2636-2642.

Burns R C and Hardy R W F 1975 Nitrogen Fixation in Bacteria and Higher Plants. Springer-Verlag. Berlin.

Bushby H V A 1982 Ecology. *In:* Nitrogen Fixation. Volume 2. *Rhizobium.* W J Broughton (ed.) pp. 35-75. Clarendon Press. Oxford.

Bushby H V A 1984 Colonization of rhizospheres and nodulation of two *Vigna* species by rhizobia inoculated onto seed: Influence of soil. Soil Biology and Biochemistry 16: 635-641.

Bushby H V A 1993 Colonization of rhizospheres by *Bradyrhizobium* spp. in relation to strain persistence and nodulation of some pasture legumes. Soil Biology and Biochemistry 25: 597-605.

Bushby H V A and Marshall K C 1977 Some factors affecting the survival of root-nodule bacteria on desiccation. Soil Biology and Biochemistry 9: 143-147.

Caballero-Mellado J and Martinez-Romero E 1999 Soil fertilization limits the genetic diversity of *Rhizobium* in bean nodules. Symbiosis 26: 111-121.

Chakrabarti S, Lee M S and Gibson A H 1981 Diversity in the nutritional requirements of strains of various *Rhizobium* species. Soil Biology and Biochemistry 13: 349-354.

Chatel D L 1969 Lupin and serradella inoculation experiments. *In:* Western Australia Department of Agriculture Plant Research Division Annual Report. 1969 - results of field experiments. p. 6. Western Australia Department of Agriculture. Perth, Australia.

Chatel D L, Greenwood D L and Parker C A 1968 Saprophytic competence as an important character in the selection of *Rhizobium* for inoculation. Transactions of the 9th International Congress of the Soil Science Society 2: 65-73.

Chatel D L and Parker C A 1973-a Survival of field-grown rhizobia over the dry summer period in Western Australia. Soil Biology and Biochemistry 5: 415-423.

Chatel D L and Parker C A 1973-b The colonization of host-root and soil by rhizobia-I. Species and strain differences in the field. Soil Biology and Biochemistry 5: 425-432.

Chowdhury M S, Marshall K C and Parker C A 1968 Growth rates of *Rhizobium trifolii* and *Rhizobium lupini* in sterilised soils. Australian Journal of Agricultural Research 19: 919-925.

Coutinho H L C, Oliveira V M, Lovato A, Maia A H N and Manfio G P 1999 Evaluation of diversity of rhizobia in Brazilian agricultural soils cultivated with soybeans. Applied Soil Ecology 13: 159-167.

Crozat Y, Cleyet-Marel J C and Corman A 1987 Use of the fluorescent antibody technique to characterize equilibrium survival concentration of *Bradyrhizobium* strains in soil. Biology and Fertility of Soils 4: 85-90.

Dakora F D and Phillips D A 1996 Diverse functions of isoflavonoids in legumes transcend antimicrobial definitions of phytoalexins. Physiological and Molecular Plant Pathology 49: 1-20.

Date R A 2001 Advances in inoculant technology: a brief review. Australian Journal of Experimental Agriculture 41: 321-325.

Date R A, Edye L A and Liu C J 1996 *Stylosanthes* sp. aff. *scabra* - a potential new forage plant for northern Australia. Tropical Grasslands 30: 133.

Denison R E 2000 Legume sanctions and the evolution of symbiotic cooperation by rhizobia. American Naturalist 156: 567-576.

Diatloff A 1977 Ecological studies of root-nodule bacteria introduced into field environments. 6. Antigenic and symbiotic stability in *Lotononis* rhizobia over a 12-year period. Soil Biology and Biochemistry 9: 85-88.

Dowling D N and Broughton W J 1986 Competition for nodulation of legumes. Annual Review of Microbiology 40: 131-157.

Dughri M H and Bottomley P J 1984 Soil acidity and the composition of an indigenous population of *Rhizobium trifolii* in nodules of different cultivars of *Trifolium subterraneum* L. Soil Biology and Biochemistry 16: 405-411.

Dye M, Skot L, Mytton L R, Harrison S P, Dooley J J and Cresswell A 1995 A study of *Rhizobium leguminosarum* biovar *trifolii* populations from soil extracts using randomly amplified polymorphic DNA profiles. Canadian Journal of Microbiology 41: 36-344.

Fillery I R P 2001 The fate of biologically fixed nitrogen in legume-based dryland farming systems: a review. Australian Journal of Experimental Agriculture 41: 361-381.

Flis S E, Glenn A R and Dilworth M J 1993 The interaction between aluminium and root nodule bacteria. Soil Biology and Biochemistry 25: 403-417.

Fuhrmann J and Wollum II A G 1989 Nodulation competition among *Bradyrhizobium japonicum* strains as influenced by rhizosphere bacteria and iron availability. Biology and Fertility of Soils 7: 108-112.

Gibson A H 1977 The influence of the environment and managerial practices on the legume-Rhizobium symbiosis. *In*: A Treatise on Dinitrogen Fixation. Section IV: Agronomy and Ecology. R W F Hardy and A H Gibson (eds.) pp. 393-450. John Wiley and Sons. New York.

Gibson A H, Date R A, Ireland J A and Brockwell J 1976 A comparison of competitiveness and persistence amongst five strains of *Rhizobium trifolii*. Soil Biology and Biochemistry 8: 395-401.

Gibson A H, Demezas D H, Gault R R, Bhuvaneswari T V and Brockwell J 1991 Genetic stability in rhizobia in the field. *In*: The Rhizosphere and Plant Growth. D L Kleister and P B Cregan (eds.) pp. 141-148. Kluwer Academic. Netherlands.

Gordon D M, Wexler M, Reardon T B and Murphy P J 1995 The genetic structure of *Rhizobium* populations. Soil Biology and Biochemistry 27: 491-499.

Grey D 1999 Short stories on the nitrogen fixation status of subterranean clover pastures in north-east Victoria. *In:* The Twelfth Australian Nitrogen Fixation Conference. The Australian Society for Nitrogen Fixation. Wagga Wagga, NSW, Australia.

Handley B A, Hedges A J and Beringer J E 1998 Importance of host plants for detecting the population diversity of *Rhizobium leguminosarum* biovar *viciae* in soil. Soil Biology and Biochemistry 30: 241-249.

Harrison S P, Young J P W and Jones D G 1989 *Rhizobium* population genetics: host preference and strain competition effects on the range of *Rhizobium leguminosarum* biovar *trifolii* genotypes isolated from natural populations. Soil Biology and Biochemistry 21: 981-986.

Hartmann A and Amarger N 1991 Genotypic diversity of an indigenous *Rhizobium meliloti* field population assessed by plasmid profiles, DNA fingerprinting, and insertion sequence typing. Canadian Journal of Microbiology 37: 600-608.

Heijnen C E, Burgers S L G E and van Veen J A 1993 Metabolic activity and population dynamics of rhizobia introduced into unamended and bentonite-amended loamy sand. Applied and Environmental Microbiology 59: 743-747.

Heijnen C E, Hok-A-Hin C H and van Veen J A 1992 Improvements to the use of bentonite clay as a protective agent, increasing survival levels of bacteria introduced into soil. Soil Biology and Biochemistry 24: 533-538.

Hennecke H, Kaluza K, Thony B, Fuhrmann M, Ludwig W and Stackebrandt E 1985 Concurrent evolution of nitrogenase genes and 16S rRNA in *Rhizobium* species and other nitrogen fixing bacteria. Archives of Microbiology 142: 342-348.

Herridge D F, Turpin J E and Robertson M J 2001 Improving nitrogen fixation of crop legumes through breeding and agronomic management: analysis with simulation modelling. Australian Journal of Experimental Agriculture 41: 391-401.

Hirsch P R 1996 Population dynamics of indigenous and genetically modified rhizobia in the field. New Phytologist 133: 159-171.

Hirsch P R and Spokes J D 1994 Survival and dispersion of genetically modified rhizobia in the field and genetic interactions with native strains. FEMS Microbiology Ecology 15: 147-160.

Howieson J G and Ewing M A 1986 Acid tolerance in the *Rhizobium meliloti-Medicago* symbiosis. Australian Journal of Agricultural Research 37: 55-64.

Howieson J G, Fillery I R P, Legocki A B, Sikorski M M, Stepkowski T, Minchin F R and Dilworth M J 1998 Nodulation, nitrogen fixation and nitrogen balance. *In:* Lupins as Crop Plants: Biology, Production and Utilization. J S Gladstones, C A Atkins and J Hamblin (eds.) pp. 149-180. University Press. Cambridge.

Howieson J G and McInnes A 2001 The legume-rhizobia symbiosis. Does it vary for the tropics relative to the Mediterranean basin? Proceedings of the XIX International Grasslands Congress, Brazil. J A Gomide, W R S Matto and S C da Silva (eds.) pp. 585-590. Brazilian Society of Animal Husbandry. Brazil.

Howieson J G, Loi A and Carr S J 1995 *Biserrula pelecinus* L. - a legume pasture species with potential for acid, duplex soils which is nodulated by unique root-nodule bacteria. Australian Journal of Agricultural Research 46: 997-1009.

Hungria M and Vargas M A T 2000 Environmental factors affecting N_2 fixation in grain legumes in the tropics, with emphasis on Brazil. Field Crops Research 65: 151-164.

Hynes M F and O'Connell M P 1990 Host plant effects on competition among strains of *Rhizobium leguminosarum*. Canadian Journal of Microbiology 36: 864-869.

Issa S and Wood M 1995 Multiplication and survival of chickpea and bean rhizobia in dry soils: the influence of strains, matric potential and soil texture. Soil Biology and Biochemistry 27: 785-792.

Jansen van Rensburg H and Strijdom B W 1985 Effectiveness of *Rhizobium* strains used in inoculants after their introduction into soil. Applied and Environmental Microbiology 49: 127-131.

Jiminez J and Casadesus J 1989 An altruistic model of the *Rhizobium*-legume association. The Journal of Heredity 80: 335-336.

Johnson H W and Means U M 1963 Serological groups of *Rhizobium japonicum* recovered from nodules of soybeans (*Glycine max*) in field soils. Agronomy Journal 55: 269-271.

Jones D G 1983 The ecology of *Rhizobium. In:* Temperate Legumes. D G Jones and D R Davies (eds.) pp. 337-371. Pitman. London.

Kay H E, Coutinho H L C, Fattori M, Manfio G P, Goodacre R, Nuti M P, Basaglia M and Beringer J E 1994 The identification of *Bradyrhizobium japonicum* strains isolated from Italian soils. Microbiology 140: 2333-2339.

Kijne J W 1992 The *Rhizobium* infection process. *In:* Biological Nitrogen Fixation. G Stacey, R H Burris and H J Evans (eds.) pp. 349-398. Chapman and Hall. New York.

Kinkle B K, Sadowsky M J, Schmidt E L and Koskinen W C 1993 Plasmids pJP4 and r68.45 can be transferred between populations of bradyrhizobia in nonsterile soil. Applied and Environmental Microbiology 59: 1762-1766.

Klingmuller W, Dally A, Fentner C and Steinlein M 1990 Plasmid transfer between soil bacteria. *In:* Bacterial Genetics in Natural Environments. J C Fry and M J Day (eds.) pp. 133-151. Chapman and Hall. London.

Ladha J K 1995 Management of biological nitrogen fixation for the development of more productive and sustainable agricultural systems. Plant and Soil 174: 1.

Laguerre G, Mavingui P, Allard M, Charnay M, Louvrier P, Mazurier S, Rigottier-Gois L and Amarger N 1996 Typing of rhizobia by PCR DNA fingerprinting and PCR-restriction fragment length polymorphism analysis of chromosomal and symbiotic gene regions: application to *Rhizobium leguminosarum* and its different biovars. Applied and Environmental Microbiology 62: 2029-2036.

Leung K and Bottomley P J 1994 Growth and nodulation characteristics of subclover (*Trifolium subterraneum* L.) and *Rhizobium leguminosarum* bv. *trifolii* at different soil water potentials. Soil Biology and Biochemistry 26: 805-812.

Leung K, Yap K, Dashti N and Bottomley P J 1994 Serological and ecological characteristics of a nodule dominant serotype from an indigenous soil population of *Rhizobium leguminosarum* bv. *trifolii.* Applied and Environmental Microbiology 60: 408-415.

Lindstrom K, Lipsanen P and Kaijalainen S 1990 Stability of markers used for identification of two *Rhizobium galegae* inoculant strains after five years in the field. Applied and Environmental Microbiology 56: 444-450.

Lochner H H, Strijdom B W and Law I J 1989 Unaltered nodulation competitiveness of a strain of *Bradyrhizobium* sp. (*Lotus*) after a decade in the soil. Applied and Environmental Microbiology 55: 3000-3008.

Louvrier P, Laguerre G and Amarger N 1996 Distribution of symbiotic genotypes in *Rhizobium leguminosarum* biovar *viciae* populations isolated directly from soils. Applied and Environmental Microbiology 62: 4202-4205.

Lowendorf H S 1980 Factors affecting survival of *Rhizobium* in the soil. Advances in Microbial Ecology 4: 87-124.

Lupwayi N Z, Stephens P M and Noonan M J 1996 Relationship between timing of infection and nodulation competitiveness of *Rhizobium meliloti.* Symbiosis 21: 233-248.

Lussenhop J 1993 Effects of two Collembola species on nodule occupancy by two *Bradyrhizobium japonicum* strains. Soil Biology and Biochemistry 25: 775-780.

Marshall K C 1964 Survival of root-nodule bacteria in dry soils exposed to high temperatures. Australian Journal of Agricultural Research 15: 273-281.

Marsudi N D S, Glenn A R and Dilworth M J 1999 Identification and characterisation of fast- and slow-growing root nodule bacteria from south-western Australian soils able to nodulate *Acacia saligna.* Soil Biology and Biochemistry 31: 1229-1238.

Martinez E, Romero D and Palacios R 1990 The *Rhizobium* genome. Critical Reviews in Plant Sciences 9: 59-93.

Martinez-Romero E 1994 Recent developments in *Rhizobium* taxonomy. Plant and Soil 161: 11-20.

Martinez-Romero E and Caballero-Mellado J 1996 *Rhizobium* phylogenies and bacterial genetic diversity. Critical Reviews in Plant Sciences 15: 113-140.

McInnes A 2002 Field Populations of Bradyrhizobia Associated with Serradella. PhD Thesis. The University of Western Australia. Perth, Australia.

Mendes I C and Bottomley P J 1998 Distribution of a population of *Rhizobium leguminosarum* bv. *trifolii* among different size classes of soil aggregates. Applied and Environmental Microbiology 64: 970-975.

Mhamdi R, Jebara M, Aouani M E, Ghrir R and Mars M 1999 Genotypic diversity and symbiotic effectiveness of rhizobia isolated from root nodules of *Phaseolus vulgaris* L. grown in Tunisian soils. Biology and Fertility of Soils 28: 313-320.

Miller M S and Pepper I L 1988 Survival of a fast growing strain of lupin rhizobia in Sonoran desert soils. Soil Biology and Biochemistry 20: 323-327.

Minamisawa K, Nakatsuka Y and Isawa T 1999 Diversity and field site variation of indigenous populations of soybean bradyrhizobia in Japan by fingerprints with repeated sequences RSα and RSβ. FEMS Microbiology Ecology 29: 171-178.

Moawad H and Beck D P 1991 Some characteristics of *Rhizobium leguminosarum* isolates from uninoculated field-grown lentil. Soil Biology and Biochemistry 23: 933-937.

Mullen M D, Israel D W and Wollum II A G 1988 Effects of *Bradyrhizobium japonicum* and soybean (*Glycine max* (L.) Merr.) phosphorus nutrition on nodulation and dinitrogen fixation. Applied and Environmental Microbiology 54: 2387-2392.

Munns D N 1977 Mineral nutrition and the legume symbiosis. *In:* A Treatise on Dinitrogen Fixation. Section IV: Agronomy and Ecology. R W F Hardy and A H Gibson (eds.) pp. 353-391. John Wiley and Sons. New York.

O'Hara G W 2001 Nutritional constraints on root nodule bacteria affecting symbiotic nitrogen fixation: a review. Australian Journal of Experimental Agriculture 41: 417-433.

O'Hara G W, Boonkerd N and Dilworth M J 1988 Mineral constraints to nitrogen fixation. Plant and Soil 108: 93-110.

O'Hara G W, Hartzook A, Bell R W and Loneragan J F 1993 Differences between *Bradyrhizobium* strains NC92 and TAL1000 in their nodulation and nitrogen fixation with peanut in iron deficient soil. Plant and Soil 155/156: 333-336.

Palmer K M and Young J P W 2000 Higher diversity of *Rhizobium leguminosarum* biovar *viciae* populations in arable soils than in grass soils. Applied and Environmental Microbiology 66: 2445-2450.

Parker C A and Oakley A E 1965 Reduced nodulation of lupins and serradella due to lime pelleting. Australian Journal of Experimental Agriculture and Animal Husbandry 5: 144-146.

Parker C A, Trinick M J and Chatel D L 1977 Rhizobia as soil and rhizosphere inhabitants. *In:* A Treatise on Dinitrogen Fixation. R W F Hardy and A H Gibson (eds.) pp. 311-352. Wiley. New York.

Paul E A and Clark F E 1996 Soil Biology and Biochemistry. Academic Press. San Diego.

Peltzer S 1996 The effect of low temperature on the growth and nodulation of *Lupinus angustifolius* (L.) PhD Thesis. The University of Western Australia. Perth, Australia.

Peoples M A and Baldock J A 2001 Nitrogen dynamics of pastures: nitrogen fixation inputs, the impact of legumes on soil nitrogen fertility, and the contributions of fixed nitrogen to Australian farming systems. Australian Journal of Experimental Agriculture 41: 327-346.

Peoples M A and Craswell E T 1992 Biological nitrogen fixation: investments, expectations and actual contributions to agriculture. Plant and Soil 141: 13-39.

Peoples M A, Herridge D F and Ladha J K 1995 Biological nitrogen fixation: an efficient source of nitrogen for sustainable agricultural production? Plant and Soil 174: 3-28.

Perez-Ramirez N O, Rogel M A, Wang E, Castellanos J Z and Martinez-Romero E 1998 Seeds of *Phaseolus vulgaris* bean carry *Rhizobium etli*. FEMS Microbiology Ecology 26: 289-296.

Phillips D A and Streit W R 1998 Modifying rhizosphere microbial communities to enhance nutrient availability in cropping systems. Field Crops Research 56: 217-221.

Povorov N A 1998 Coevolution of rhizobia with legumes: facts and hypotheses. Symbiosis 24: 337-368.

Povorov N A and Vorob'ev N I 1998 The role of interstrain competition in the evolution of genetically polymorphic populations of nodule bacteria. Genetika 34: 1712-1719.

Robinson A C 1969-a Competition between effective and ineffective strains of *Rhizobium trifolii* in the nodulation of *Trifolium subterraneum*. Australian Journal of Agricultural Research 20: 827-841.

Robinson A C 1969-b Host selection for effective *Rhizobium trifolii* by red clover and subterranean clover in the field. Australian Journal of Agricultural Research 20: 1053-1060.

Rodriguez-Navarro D N, Buendia A M, Camacho M, Lucas M M and Santamaria C 2000 Characterisation of *Rhizobium* spp. bean isolates from south-west Spain. Soil Biology and Biochemistry 32: 1601-1613.

Ronson C W 1999 Application of molecular biology to improving *Rhizobium* strains. *In*: The Twelfth Australian Nitrogen Fixation Conference. The Australian Society for Nitrogen Fixation. Wagga Wagga, NSW, Australia.

Roughley R J, Blowes W M and Herridge D F 1976 Nodulation of *Trifolium subterraneum* by introduced rhizobia in competition with naturalized strains. Soil Biology and Biochemistry 8: 403-407.

Rutherford P M and Juma N G 1992 Influence of texture on habitable pore space and bacterial-protozoan populations in soil. Biology and Fertility of Soils 12: 221-227.

Saito A, Mitsui H, Hattori R, Minamisawa K and Hattori T 1998 Slow-growing and oligotrophic soil bacteria phylogenetically close to *Bradyrhizobium japonicum*. FEMS Microbiology Ecology 25: 277-286.

Sanginga N, Danso S K A, Mulongoy K and Ojeifo A A 1994 Persistence and recovery of introduced *Rhizobium* ten years after inoculation of *Leucaena leucocephala* grown on an Alfisol in southwestern Nigeria. Plant and Soil 159: 199-204.

Santos M A, Vargas M A T and Hungria M 1999 Characterisation of soybean *Bradyrhizobium* strains adapted to Brazilian savannas. FEMS Microbiology Ecology 30: 261-267.

Sattar M A, Quader M A and Danso S K A 1995 Nodulation, N_2 fixation and yield of chickpea as influenced by host cultivar and *Bradyrhizobium* strain differences. Soil Biology and Biochemistry 27: 725-727.

Schmidt E L 1988 Competition for legume nodule occupancy: a down-to-earth limitation on nitrogen fixation. *In*: World Crops: Cool Season Food Legumes. R J Summerfield (ed.) pp. 663-674. Kluwer Academic Publishers. Netherlands.

Segovia L, Pinero D, Palacios R, and Marintez-Romero E 1991 Genetic structure of a soil population of nonsymbiotic *Rhizobium leguminosarum*. Applied and Environmental Microbiology 57: 426-433.

Selander R K, Caugant D A, Ochman H, Musser J M, Gilmour M N and Whittam T S 1986 Methods of multilocus enzyme electrophoresis for bacterial population genetics and systematics. Applied and Environmental Microbiology 51: 873-884.

Silva C, Eguiarte L E and Souza V 1999 Reticulated and epidemic population genetic structure of *Rhizobium etli* biovar *phaseoli* in a traditionally managed locality in Mexico. Molecular Ecology 8: 277-287.

Singleton P W, El-Swaify S A and Bohlool B B 1982 Effect of salinity on *Rhizobium* growth and survival. Applied and Environmental Microbiology 44: 884-890.

Slattery J F and Coventry D R 1993 Variation of soil populations of *Rhizobium leguminosarum* bv. *trifolii* and the occurrence of inoculant rhizobia in nodules of subterranean clover after pasture renovation in north-eastern Victoria. Soil Biology and Biochemistry 25: 1725-1730.

Slattery J F, Coventry D R and Slattery W J 2001 Rhizobial ecology as affected by the soil environment. Australian Journal of Experimental Agriculture 41: 289-298.

Souza V, Eguiarte L, Avila G, Cappello R, Gallardo C, Montoya J and Pinero D 1994 Genetic structure of *Rhizobium etli* biovar *phaseoli* associated with wild and cultivated bean plants (*Phaseolus vulgaris* and *Phaseolus coccineus*) in Morelos, Mexico. Applied and Environmental Microbiology 60: 1260-1268.

Sprent J I 1994 Evolution and diversity in the legume-*Rhizobium* symbiosis: chaos theory? Plant and Soil 161: 1-10.

Sprent J I 2001 Nodulation in Legumes. Royal Botanic Gardens, Kew, UK.

Strain S R, Leung K, Whittam T S, de Bruijn F J and Bottomley P J 1994 Genetic structure of *Rhizobium leguminosarum* biovar *trifolii* and *viciae* populations found in two Oregon soils under different plant communities. Applied and Environmental Microbiology 60: 2772-2778.

Strain S R, Whittam T S and Bottomley P J 1995 Analysis of genetic structure in soil populations of *Rhizobium leguminosarum* recovered from the USA and UK. Molecular Ecology 4: 105-114.

Sullivan J T, Eardly B D, van Berkum P and Ronson C W 1996 Four unnamed species of nonsymbioticrhizobia isolated from the rhizosphere of *Lotus corniculatus*. Applied and Environmental Microbiology 62: 2818-2825.

Sullivan J T, Patrick H N, Lowther W L, Scott D B and Ronson C W 1995 Nodulating strains of *Rhizobium loti* arise through chromosomal symbiotic gene transfer in the environment. Proceedings of the National Academy of Science USA 92: 8985-8989.

Sullivan J T and Ronson C W 1998 Evolution of rhizobia by acquisition of a 500-kb symbiosis island that integrates into a phe-tRNA gene. Proceedings of the National Academy of Sciences USA 95: 5145-5149.

Tang C, Robson A D, Dilworth M J and Kuo J 1992 Microscopic evidence on how iron deficiency limits nodule initiation in *Lupinus angustifolius* L. New Phytologist 121: 457-467.

Thies J E, Bohlool B B and Singleton P W 1992 Environmental effects on competition for nodule occupancy between introduced and indigenous rhizobia and among introduced strains. Canadian Journal of Microbiology 38: 493-500.

Thies J E, Singleton P W and Bohlool B B 1991 Influence of size of indigenous rhizobial populations on establishment and symbiotic performance of introduced rhizobia on field-grown legumes. Applied and Environmental Microbiology 57: 19-28.

Thies J E, Woomer P L and Singleton P W 1995 Enrichment of *Bradyrhizobium* spp. populations in soil due to cropping of the homologous host legume. Soil Biology and Biochemistry 27: 633-636.

Thurman N P and Bromfield E S P 1988 Effect of variation within and between *Medicago* and *Melilotus* species on the composition and dynamics of indigenous populations of *Rhizobium meliloti*. Soil Biology and Biochemistry 20: 31-38.

Trevors J T, Barkay T and Bourquin A W 1987 Gene transfer among bacteria in soil and aquatic environments: a review. Canadian Journal of Microbiology 33: 191-198.

Trinick M J 1985 *Rhizobium* strain competition for host nodulation. *In*: World Soybean Research Conference III. R Shibles (ed.) pp. 911-917. Westview Press. Boulder.

Trinick M J, Parker C A and Palmer M J 1983 Interactions of the microflora from nodulation problem and non-problem soils towards *Rhizobium* spp. on agar culture. Soil Biology and Biochemistry 15: 295-301.

Turco R F and Bezdicek D F 1987 Diversity within two serogroups of *Rhizobium leguminosarum* native to soils in the Palouse of eastern Washington. Annals of Applied Biology 111: 103-114.

Udvardi M K and Khan M L 1992 Evolution of the (*Brady*) *Rhizobium* - legume symbiosis: why do bacteroids fix nitrogen? Symbiosis 14: 87-101.

Unkovich M and Pate J S 1998 Symbiotic effectiveness and tolerance to early season nitrate in indigenous populations of subterranean clover rhizobia from SW Australian pastures. Soil Biology and Biochemistry 30: 1435-1443.

Unkovich M, Pate J S, Armstrong E L and Sanford P 1995 Nitrogen economy of annual crop and pasture legumes in southwest Australia. Soil Biology and Biochemistry 27: 585-588.

van Elsas J D and Trevors J T 1990 Plasmid transfer to indigenous bacteria in soil and rhizosphere: problems and perspectives. *In*: Bacterial Genetics in Natural Environments. J C Fry and M J Day (eds.) Chapman and Hall. London.

Vincent J M 1965 Environmental factors in the fixation of nitrogen by the legume. *In*: Soil Nitrogen. W V Bartholomew and F E Clark (eds.) pp. 384-435. American Society of Agronomy. Madison. USA.

Weaver R W, Sen D, Coll J J, Dixon C R and Smith G R 1989 Specificity of arrowleaf clover for rhizobia and its establishment on soil from crimson clover pastures. Soil Science Society of America Journal 53: 731-734.

Weber D F, Keyser H H and Uratsu S L 1989 Serological distribution of *Bradyrhizobium japonicum* from U.S. soybean production areas. Agronomy Journal 81: 786-789.

Wolff A B, Streit W, Kipe-Nolt J A, Vargas H and Werner D 1991 Competitiveness of *Rhizobium leguminosarum* bv. *phaseoli* strains in relation to environmental stress and plant defence mechanisms. Biology and Fertility of Soils 12: 170-176.

Woomer P, Asano W and Pradhan D 1990 Environmental factors related to rhizobial abundance in kikuyugrass (*Pennisetum clandestinum*) pastures. Tropical Agriculture (Trinidad) 67: 217-220.

Woomer P and Bohlool B B 1989 Rhizobial ecology in tropical pasture legumes. *In*: Persistence in Forage Legumes. G C Marten, A G Matches, R F Barnes, R W

Brougham, R J Clements and G W Sheath (eds.) pp. 233-245. American Society of Agronomy. Madison.

Woomer P, Singleton P W and Bohlool B B 1988 Ecological indicators of native rhizobia in tropical soils. Applied and Environmental Microbiology 54: 1112-1116.

Woomer P L 1990 Predicting the abundance of indigenous and the persistence of introduced rhizobia in tropical soils. PhD Thesis. The University of Hawaii. Paia.

Young J P W 1985 *Rhizobium* population genetics: enzyme polymorphism in isolates from peas, clover, beans and lucerne grown at the same site. Journal of General Microbiology 131: 2399-2408.

Young J P W 1990 Molecular population genetics and evolution of rhizobia. *In*: The Nitrogen Fixation and its Research in China. G F Hong (ed.) pp. 365-381. Springer-Verlag. Berlin.

Young J P W 1993 Molecular phylogeny of rhizobia and their relatives. *In*: New Horizons in Nitrogen Fixation. R Palacios, J Mora and W E Newton (eds.) pp. 587-592. Kluwer. Netherlands.

Young J P W, Demetriou L and Apte R G 1987 *Rhizobium* population genetics: enzyme polymorphism in *Rhizobium leguminosarum* from plants and soil in a pea crop. Applied and Environmental Microbiology 53: 397-402.

Young J P W and Haukka K E 1996 Diversity and phylogeny of rhizobia. New Phytologist 133: 87-94.

Young J P W and Wexler M 1988 Sym plasmid and chromosomal genotypes are correlated in field populations of *Rhizobium leguminosarum*. Journal of General Microbiology 134: 2731-2740.

Chapter 7

Contributions of Arbuscular Mycorrhizas to Soil Biological Fertility

David D. Douds, Jr.[1] and Nancy Collins Johnson[2]

[1] *USDA-ARS, Eastern Regional Research Center, 600 E. Mermaid Lane, Wyndmoor, PA 19038, USA.*

[2] *Northern Arizona University, Environment Sciences and Biological Sciences, P.O. Box 5694, Flagstaff, AZ 86001, USA.*

1. INTRODUCTION

Mycorrhizas are ubiquitous plant-fungal associations that are important components of soil fertility (Table 1). Roots of most crops are normally inhabited by arbuscular mycorrhizal (AM) fungi. These Zygomycota in the order Glomales, function at the interface between plants and soils by greatly expanding the area from which plants can gather soil resources. Extensive networks of as much as 160 m of AM hyphae per g of soil (Degens *et al.* 1994) function as conduits for nutrient uptake. Crops with coarse root systems generally benefit greatly from AM associations, while mycorrhizal benefits in crops with more fibrous root systems tend to be determined by soil mineral availability (Baylis 1975, Hetrick *et al.* 1992). Only a few crops, such as lupines and members of the Brassicaceae and Chenopodiaceae, do not regularly form AM associations. In addition to their direct effects on nutrient uptake, AM fungi also contribute to soil fertility by enhancing soil structure and protecting crops from root pathogens.

L.K. Abbott & D.V. Murphy (eds) *Soil Biological Fertility - A Key to Sustainable Land Use in Agriculture.* 129-162.

AM fungi form structures inside (intraradical) and outside (extraradical) the host root. After an infective soil borne hypha contacts a host root, it forms an appressorium, penetrates the epidermis, and grows in the space between the cells into the root cortex. Once in the cortex, the intraradical hyphae penetrate the cells and produce arbuscules. These highly branched structures are surrounded by the host cells' membranes and are thought to be where nutrients are exchanged between the partners: i.e. glucose from host to fungus and phosphorus (P) from fungus to host (Blee and Anderson 1998). Once the source of carbohydrate nourishment is secured, the extraradical hyphae can proliferate in the soil. New spores are produced typically on hyphae in the soil in response to achievement of a critical amount of root length colonised, senescence of the host, or other factors. There are clear differences in the effectiveness of AM fungal species to improve soil fertility (Abbott and Robson 1982, 1985, Graham and Abbott 2000), and these differences are likely to be related to differences in allocation to intraradical and extraradical structures (Abbott and Gazey 1994, Dodd *et al.* 2000).

Table 1 Arbuscular mycorrhizal functions that can ameliorate soil fertility*

Direct **positive** effects upon:	Lessen **negative** effects from:
• Uptake of immobile nutrients • Drought tolerance • Soil macroaggregate formation and stability • Soil organic matter	• Root pathogens • Leaching loss of nutrients • Microbial immobilisation of nutrients

*Combinations of crops, AM fungi, and soils differ greatly in their function. Most AM associations do not simultaneously enhance all of these components of soil fertility, and some may not enhance any of them.

The partnership between plants and AM fungi has a long history. Fossil and molecular evidence indicates that AM fungi were associated with the earliest land plants, and that the symbiosis evolved concurrently with the evolution of roots (Malloch *et al.* 1980, Stubblefield *et al.* 1987, Simon *et al.* 1993, Redecker *et al.* 2000). The intimacy of this association is reflected in the fact that Glomalean fungi are obligate biotrophs that have not yet been successfully cultured in the absence of root tissues. Although the Glomales are asexual and include fewer than 160 species (INVAM 2001), a surprising level of genetic diversity is maintained within populations of these fungi (Hijri *et al.* 1999, Hosny *et al.* 1999). Sanders (1999) suggested that, in genetic terms, an individual aseptate AM fungus is actually a population of discrete nuclei. This genetic variance within taxa corroborates physiological variance between geographic isolates of the same species. For example, Bethlenfalvay *et al.* (1989) found that *Glomus mosseae* isolated from an arid site improved the photosynthetic water use efficiency of soybean more than *G. mosseae* isolated from a mesic site. Other studies also have shown that different isolates of the same species can elicit different plant responses under identical conditions (e.g. Stahl and Smith 1984, Stahl and Christensen 1990, Sylvia *et al.* 1993-a).

Mycorrhizal function is strongly influenced by the soil environment, particularly those factors that control mineral fertility (Abbott and Robson 1982). Generally, mycorrhizal benefits are greater in phosphorus-poor soils than in phosphorus-rich ones (Koide 1991). Furthermore, crop species and even different cultivars of the same species interact with AM fungi differently (Hetrick *et al.* 1993, Hetrick *et al.* 1996). It is useful to envision mycorrhizas as dynamic systems controlled by interactions among plants, fungi, soil microbes, and soil properties. Bethlenfalvay and Schüepp (1994) suggested that sustainable agroecosystems require management to generate a stable community of soil biota that functions effectively with abiotic conditions to maximise crop productivity and minimise inputs and soil erosion.

Cultural practices have been shown to influence the species composition of AM fungal communities (see below). Certain taxa increase in abundance in agricultural systems relative to other taxa. Furthermore, species diversity of AM fungi is consistently lower in agricultural systems than in nearby natural areas (Sieverding 1990, Helgason *et al.* 1998-b). The consequences of this on crop production have not yet been carefully studied. Sieverding (1990) suggested that a few well selected AM fungi could increase yields if they are the best mutualists. Alternatively, if the proliferating fungi are simply the most aggressive colonists, and not the best at improving nutrient uptake, pathogen resistance, or soil structure, then this agriculture-induced reduction in diversity is cause for concern.

The potential for agricultural management of mycorrhizas to reduce reliance on inorganic fertilisers and develop more sustainable agricultural systems has long been recognised and has already been reviewed (e.g. Sanders *et al.* 1975, Azcon-Aguilar *et al.* 1979, Bethlenfalvay and Linderman 1992, Pfleger and Linderman 1994, Gianinazzi and Schüepp 1994). But, "promises of the applied value of AM fungi in agriculture, forestry and horticulture have been more rhetorical than deliverable" (Miller and Jastrow 1992). A much better understanding of the ecological and evolutionary mechanisms responsible for generating positive, neutral or negative mycorrhizal functioning in field environments is necessary before mycorrhizas can be effectively managed to maximise their contribution to soil fertility in sustainable systems.

This chapter has a twofold emphasis. First, it describes the fundamental ways in which AM fungi contribute to the biological fertility of the soil (Table 1). We discuss how AM fungi directly affect plant growth and soil structure and how their interactions with other soil organisms indirectly affect crop yields and nutrient cycling. We will see that these are not independent effects and that feedbacks between plants, fungi, and biotic and abiotic soil properties ultimately determine mycorrhizal effects on plant growth. Second, we discuss how agricultural management practices affect indigenous communities of these fungi. We show that management practices positively and negatively affect AM fungi, and that these have ramifications upon plant growth. Throughout we will point out key topics where further research is needed.

2.   GENERAL IMPACTS OF ARBUSCULAR MYCORRHIZAS

2.1   Plant Growth

Arbuscular mycorrhizal fungi long have been known to have a positive effect upon growth of their host plant (Mosse 1973), most notably in low nutrient soils. This is due to enhanced nutrient uptake, water relations, and disease resistance. These benefits are often contingent on environmental conditions, and when nutrients and water are in unlimited supply and pathogens are absent, then the costs of AM symbioses may sometimes outweigh their benefits and AM fungi may actually depress plant growth (Fitter 1991, Johnson *et al.* 1997).

2.1.1   Nutrient uptake

The extraradical phase of the mycorrhiza acts in effect as an extension of the root system for the uptake of nutrients, particularly those which are relatively immobile in the soil solution, i.e. phosphate, zinc (Zn), and copper (Cu). The zone of P uptake from the soil for a nonmycorrhizal root extends only just beyond the length of a root hair, 1-2 mm in most instances (Jungk and Claassen 1986). Hyphae of AM fungi can extend upwards of 14 cm beyond the root (Mozafar *et al.* 2001), effectively exploring a greater volume of soil for nutrients. This phenomenon was demonstrated clearly in experiments utilising compartmented pots (Li *et al.* 1991). Plants were grown in pots separated into root and hyphal compartments by a screen with mesh size small enough to restrict passage of roots yet allow penetration by hyphae. These experiments have shown a zone of uptake of soil P extending through the entire hyphal compartment for mycorrhizal plants and no uptake from this compartment by nonmycorrhizal plants.

There was some debate about whether AM fungi make available to the plant, or solubilise, unavailable forms of P such as rock phosphate. Much of the belief that AM fungi could solubilize unavailable forms of P came from the observation that mycorrhizal plants were more efficient at obtaining P than nonmycorrhizal plants in the presence of insoluble rock phosphate fertiliser (Powell and Daniel 1978). However, experiments with ^{32}P labelled fertilisers (Hayman and Mosse 1972), and others (Bolan 1991, Nurlaeney *et al.* 1996), indicated uptake of P only from the available pool. AM fungi may enhance plant uptake of rock P by a Le Chatelier's Principle type of mechanism. As P is taken up from the soil solution by the hyphae, more P enters the soil solution from sparsely-soluble forms of P (Ness and Vlek 2000).

AM fungi may allow plants to better utilise organic forms of P in the soil. The extraradical hyphae of mycorrhizas have phosphatase activity associated with their cell walls (Joner and Johansen 2000). Hydrolysis of organic P by extraradical hyphae and transport of that P to host roots recently was demonstrated *in vitro* (Joner *et al.* 2000).

Plants may limit colonisation of their roots in soils of high P availability (Menge *et al.* 1978). This serves to limit the carbon cost of supplying the metabolic needs of the fungus, which may be substantial, ranging from 4 to 20 % of plant photosynthate in the absence of enhanced nutrient uptake (Graham 2000, Douds *et al.* 2000).

2.1.2 Water balance

Another way in which AM fungi affect the growth of their host plant is through enhanced water balance by altering the behaviour of their stomata. Increased stomatal conductance, and hence transpiration, has been noted in mycorrhizal compared to nonmycorrhizal plants under both well-watered and drought conditions (see review; Augé 2000). P-supplemented nonmycorrhizal plants often function as do mycorrhizal plants under these conditions but transpiration and/or stomatal conductances have been measured to be greater in mycorrhizal than nonmycorrhizal plants when both groups were of similar size (Bryla and Duniway 1998, Allen 1982, Augé 2000) and when leaf water potentials were similar (Allen *et al.* 1981, Augé *et al.* 1986).

The mechanisms whereby arbuscular mycorrhizas enhance the water balance of their hosts is a matter of debate. When cultural conditions lead to a growth or P nutritional response to mycorrhizas (i.e. larger plants or increased P status relative to controls), reasons for enhanced stomatal conductance of mycorrhizal plants may be more apparent. Larger root systems access water from a greater volume of soil. Increased P in leaves allows for more rapid export of photosynthates so stomates remain open longer (Jarvis and Davies 1998). However, other factors must be operating when plants of similar size and P status are compared. Actual water uptake and its movement to the root by hyphae of AM fungi may (Ruiz-Lozano and Azcon 1995) or may not (George *et al.* 1992) occur. The enhanced water relations of mycorrhizal plants may be due to the effect of hyphae upon soil structure (see below) and the resulting influence upon water holding properties. Another way mycorrhizas may influence stomatal conductance is alteration of non-hydraulic root-to-shoot signalling of soil drying (Ebel *et al.* 1996, Augé and Duan 1991), keeping stomata open longer as portions of a root system are exposed to dry soil.

2.1.3   Resistance to plant diseases and pests

Mycorrhizas also confer upon their hosts a measure of resistance to a variety of soil borne diseases and pests (Table 2). As was found with water relations, some of these instances of disease resistance were due to enhanced nutrition of the mycorrhizal plant. Increased nutrition of nonmycorrhizal plants and the enhanced vigour it causes can lead to disease resistance (Graham and Egel 1988). Other root-pathogen interactions affected by the mycorrhiza require pre-colonisation of the roots by AM fungi prior to challenge by the pathogen (Afek *et al.* 1990). This suggests that AM fungi and the pathogen may compete for host derived carbon and/or infection sites, or that pre-colonisation of the root system potentiates the host defence system (Benhamou *et al.* 1994).

Another mechanism, not usually considered in these studies, has been proposed by Linderman (2000). In addition to the rhizosphere, the volume of soil directly influenced by the root, one may also consider a "mycorrhizosphere," the volume of soil influenced by the extraradical phase of the mycorrhiza. Just as roots influence the rhizosphere microflora through exudation, sloughing of cells, and root turnover, mycorrhizal hyphae influence the microflora of the mycorrhizosphere (see above). In addition, their influence is amplified through mycorrhiza-mediated changes in root exudation (Norman and Hooker 2000). These changes in root exudation, or exudation and other influences on soil chemistry by the hyphae themselves, may directly affect pathogens or other soil microbes (Filion *et al.* 1999). These other soil fungi and bacteria, influenced by the mycorrhiza, can be antagonistic to pathogens. Linderman (2000) has measured the "antagonistic potential" of bacteria isolated from the rhizosphere and mycorrhizosphere of nonmycorrhizal and mycorrhizal plants against a variety of plant pathogens. Antagonistic potential is a measure of the zones of inhibition around bacteria colonies isolated from these "spheres," when challenged by pathogenic fungi. Bacterial isolates from mycorrhizosphere soils were more antagonistic to plant pathogenic fungi than those from the rhizospheres of nonmycorrhizal roots (Linderman, 2000).

Table 2 Demonstrated resistance to fungal diseases conferred to the host plant by AM fungi

Pathogen	Host plant	AM fungus	Reference
Fusarium oxysporum	*Lycopersicon esculentum*	*Glomus Intraradices*	Caron *et al.* 1986
	Asparagus officinale	*Glomus fasciculatum*	Wacker *et al.* 1990
	Daucus carota	*G. intraradices*	Benhamou *et al.* 1994
	Vulpia ciliata	*Glomus sp.*	Newsham *et al.* 1995
Thielaviopsis brasicola	*Nicotiana tobaccum*	*Glomus monosporum*	Giovannetti *et al.* 1991
Pythium ultimum	*Tagetes patula*	*G. intraradices*	St-Arnaud *et al.* 1994
Verticillium dahliae	*Solanum melongena*	*Glomus etunicatum*	Matsubara *et al.* 1995.
		Gigaspora margarita	Matsubara *et al.* 1995.
	Gossypium hirsutum	*Glomus versiforme*	Liu 1995
Cylindrocarpon destructans	*Prunus persica*	*Glomus aggregatum*	Traquair 1995
Phytophthora nicotianae	*L. esculentum*	*Glomus mosseae*	Cordier *et al.* 1996 Trotta *et al.* 1996
P. parasitica	*L. esculentum*	*G. mosseae*	Cordier *et al.* 1996 Vigo *et al.* 2000
P. fragariae	*Fragaria X ananassa*	*G. etunicatum* *G. monosporum*	Norman and Hooker 2000
Sclerotium cepivorum	*A. cepa*	*Glomus sp.*	Torres-Barragan *et al.* 1996
Aphanomyces eutreiches	*Pisum sativum*	*G. mosseae*	Slezack *et al.* 2000
Fusarium solani	*Phaseolus vulgarus*	*G. mosseae*	Dar *et al.* 1997

Other situations in which field functioning of AM fungi has been demonstrated bear discussion. Large scale flooding occurred along the Mississippi River in 1993. When floodwaters finally receded, maize grown in these areas the following year was stunted and exhibited P deficiency despite adequate soil test P levels (Wetterauer and Killorn 1996, Ellis 1998). Assays showed low levels of AM fungus colonisation of roots, and supplemental P fertilisation eliminated the P deficiency. The reduced levels of inoculum of AM fungi was due more to the extended fallow rather than the flooding *per se* (Ellis 1998), reminiscent of long fallow disorder (Thompson 1987, Thompson 1991, see below).

2.2   Effect of AM Fungi on Soil Structure

The organisation of soil particles into macroaggregates is important for soil aeration, water infiltration, resistance to erosion, and hence, is also important for plant growth. Tisdall and Oades (1982) proposed a hierarchical theory for the formation of soil macroaggregates. According to this theory, microaggregates (0.02- 0.25 mm in diameter) are formed from electrostatic interactions of primary clay particles and organic matter. These structures are highly stable in soil (Tisdall 1991). Macroaggregates form from microaggregates by processes that are not fully understood (Degens *et al*. 1994).

AM fungi are believed to play a role in the stabilisation of microaggregates into macroaggregates (Miller and Jastrow 2000). A number of studies have correlated the presence of mycorrhizas with increased water stable macroaggregates (Schreiner *et al*. 1997, Thomas *et al*. 1986, Miller and Jastrow 1990) though the effects of the fungus are difficult to dissociate from those of the root. Thomas *et al*. (1993) used split-root plants growing in four-chambered pots in a silty clay loam soil and compared water stable soil aggregates in soils containing all combinations (presence/absence) of roots and extraradical AM fungus hyphae. Though aggregation was greatest in the mycorrhizal root chamber, there were similar percentages of water stable aggregates in the nonmycorrhizal root *vs* the hyphae-only chamber. They concluded that the root and hyphae have similar effects on the stability of soil aggregates. Miller and Jastrow (1990) studied mycorrhizas in a chronosequence of tallgrass prairie restoration on a silt loam soil in Illinois, USA. They have used path analysis to quantify the relative contributions of extraradical hyphae, fine and very fine roots, and various soil organic matter pools to the formation of stable soil macroaggregates (Miller and Jastrow 1990, Jastrow *et al*. 1998). This analysis showed that the hyphae had a greater direct role in stabilising the aggregates than did fine or very fine roots, but the indirect effect of very fine roots, through their symbiosis with AM fungi, was substantial.

These studies support the view of the hyphae stabilising soil particles through a mechanism of physical entanglement. Indeed, the amount of hyphae calculated to be present in soil aggregates is impressive. Each gram of stable macroaggregates can contain 50-160 m of hyphae (Tisdall and Oades 1979, Degens *et al.* 1994). Microscopic examination also has allowed the visualisation of this phenomenon (see refs in Degens *et al.* 1994).

A key mechanism of AM fungus stabilisation of soil aggregates appears to be an iron containing glycoprotein termed "glomalin" (Wright *et al.* 1996, Wright and Upadhyaya 1998). Aggregate stability and glomalin content of soils have been positively correlated (Wright and Upadhyaya 1998, Wright *et al.* 1999, Wright and Anderson 2000). Soils may contain 4.4 to 14.8 mg glomalin per g. This hydrophobic molecule is produced by all AM fungi examined and is deposited on the walls of extraradical hyphae. Evidence suggests that *Gigaspora* spp produce more glomalin per mg hyphae than do *Glomus* spp (Wright and Upadhyaya 1996). This supports findings by Miller and Jastrow (1992) who found that one species in particular, *Gigaspora gigantea*, was most associated with macroaggregation of soil in the tallgrass prairie restoration chronosequence. Immuno-fluorescent assays have demonstrated its appearance on roots and root hairs of mycorrhizal plants, AM fungal spores (Wright and Upadhyaya 1996) and soil aggregates (Wright and Upadhyaya 1998). Further, glomalin and water stable soil aggregates are linked with agricultural management practices. The transition in tillage from ploughing to no-till increased both water stable aggregates and soil glomalin (Wright *et al.* 1999). Aggregate stability varies with glomalin in soils under various crop rotations (Wright and Anderson 2000).

There is a definite need for the involvement of other disciplines in the study of glomalin and its role in soil aggregation. Soil chemists and physicists should study how this glycoprotein interacts with soil particles and why it is so recalcitrant. Since this compound can be 1% of the weight of upper layers of soil, it represents a significant portion of total soil organic matter and as such deserves further study. Very little is known about its structure, and nothing is known about its biosynthesis and secretion outside the hyphal wall. Also, the impact of plant and soil P status on the production of glomalin is not known. This is important with the increasing application of P-rich animal manures on many soils.

Most studies linking AM fungi and other biological processes to soil aggregation have been conducted in fine textured soils. There is some doubt as to whether AM fungi play a role in soil aggregation in sandy soils however (Degens *et al.* 1994). Stabilisation of aggregates in these soils may be limited by the inhibitory effect of large particle sizes upon aggregation.

Lengths of hyphae well in excess of 50 m per g of aggregates may be needed for AM fungus hyphae to contribute to water stability of aggregates (Degens *et al.* 1994). In addition, the effect of an AM fungus upon aggregation may differ for different soil types. An isolate of *Glomus mosseae* improved soil aggregation by 400% in a gray silt-loam high in organic matter and P, but in a yellow clay-loam low in organic matter and P the same fungus had a much smaller (50%) affect on aggregate stability (Bethlenfalvay and Barea 1994).

2.3   Interactions among AM Fungi and Other Soil Organisms

Increasing attention is being paid to the complex interactions among AM fungi and other soil organisms because these relationships can potentially enhance or eliminate mycorrhizal benefits for crop production and soil stabilisation (Bethlenfalvay and Schüepp 1994, Hodge 2000). As mentioned previously, AM fungal colonisation changes the chemistry of roots and exudates and generates a 'mycorrhizosphere community' of microorganisms that is distinct from that of the rhizosphere of nonmycorrhizal roots (Linderman 2000). Furthermore, because AM fungal species, and even isolates of the same species, differ in their influence on roots and exudates, microbial assemblages differ in the mycorrhizospheres of different AM fungal isolates (Meyer and Linderman 1986, Schreiner *et al.* 1997, Andrade *et al.* 1997). The activities of soil bacteria, actinomycetes, fungi, mites, collembolan and nematodes can influence the formation and functioning of mycorrhizal associations through a variety of mechanisms (Table 3). This finding opens the possibility that mycorrhizal function may result from a consortium of soil organisms that are associated with AM fungi and not from the fungi alone (Bethlenfalvay and Schüepp 1994, Gryndler 2000).

Table 3 Mechanisms by which biotic interactions can mediate mycorrhizal function

Biotic interactions mediate mycorrhizal function through:

- Changing the availability of essential resources/substrates
- Producing stimulatory or inhibitory compounds
- Modifying rhizosphere chemistry
- Grazing extraradical hyphal networks
- Modifying soil structure
- Dispersing or destroying propagules

Soil organisms can be either beneficial or antagonistic to AM fungi. Nearly forty years ago certain bacteria were shown to enhance germination of AM fungus spores (Mosse 1962), and since that time, many other beneficial interactions between AM fungi and bacteria have been observed. Garbaye (1994) reviewed the scope of these associations and defined 'Mycorrhization Helper Bacteria' (MHB) as "bacteria associated with mycorrhizal roots and mycorrhizal fungi which selectively promote the establishment of mycorrhizal symbiosis." Beneficial associations can also be mediated through the host plant. For example, by reciprocally supplying P and nitrogen (N) to a common plant host, AM fungi and N-fixing bacteria generate a synergistic tripartite symbiosis which is superior to a dual symbiosis, with either the AM fungus or diazotroph individually (Barea *et al.* 1992, Biro *et al.* 2000). Antagonists of AM fungi include mycoparasites, spore and hyphal grazers, and competitors. Detrimental effects of antagonistic soil organisms on AM fungi and their hosts have been recognised for over twenty years (e.g. Ross and Ruttencutter 1977, Ross 1980, Wilson *et al.* 1988), but much work remains before the mechanisms of these interactions are understood. As the natural history of associations between AM fungi and other soil organisms becomes better elucidated, it will be possible to design management strategies that deter organisms that are antagonists of mutualistic AM fungi and stimulate organisms that are beneficial to them.

Although field-based research is necessary to develop management strategies that maximise the beneficial AM fungus-microbe interactions and minimise the detrimental ones, to date, studies of AM fungus interactions with soil microbes have largely been confined to pots in glasshouses or growth chambers. This is because the staggering diversity and rapid growth rates of most soil organisms in the field often makes field studies of these interactions too complicated for human comprehension. One way microbial ecologists study tremendously diverse microbial communities is to make generalisations from 'functional groups' of microbes. Functional groups have been defined in various ways, usually according to tropic status or specific physiological requirements. Nutritional profiles of components of communities of soil microbes are now routine using standardised carbon sources, such as in Biolog (Biolog Inc. Hayward, CA) microplates (Garland and Mills 1991). Future mycorrhizal research may make significant advances using Biolog microplates designed to reflect the availability of carbon substrates in mycorrhizospheres vs uncolonized soil. For example, one of the few carbon substrates known to be taken up by extraradical hyphae of AM fungi is acetate (Bago *et al.* 2000). One would therefore expect limited availability of acetate in the soil of the mycorrhizosphere *vs* the bulk soil. This may affect the microbial community. Enzyme assays are another technique that could be used in field-based research of AM-soil microbe interactions (Sinsabaugh 1994). This technique quantifies extracellular enzymes and reflects the actual physiological activity of microbes and could be used to describe the differences in enzymatic activities of soil microbial communities as affected by AM fungi or various management practices which affect AM fungi. Both of these methods have great value in community level exploration. Once interactions among functional groups of organisms are identified, PCR, DNA probes, and other molecular or immunological tools can be applied to track the organisms involved and better understand the mechanisms of the interactions (Table 3). The following discussion briefly summarises the range of feedbacks that exist between soil organisms and begins to identify the kinds of interactions that will most likely lead to sustainable mycorrhizal benefits.

2.3.1 Resource availability

Transfer of essential resources is a strong mediator of species interactions. A resource can be defined as any substance that is consumed by an organism and can lead to increased growth rates as its availability in the environment increases (Tilman 1988). According to this definition, the copious extraradical hyphae produced by AM fungi provide substrates for soil microbes. Because extraradical AM hyphae can transport significant quantities of carbon substrate into the soil (Jakobsen and Rosendahl 1990), one might expect that total populations of soil organisms should consistently be elevated in mycorrhizosphere soils compared to rhizosphere soils of nonmycorrhizal plants. However, this is not the case, total microbial populations are often lower in the mycorrhizosphere (Ames *et al.* 1984, Christenson and Jakobsen 1993, Andrade *et al.* 1997, 1998). This suggests that AM fungi and certain soil microbes compete for the same rhizosphere substrates. As mentioned above, this is likely to be an important mechanism by which AM fungi protect their hosts from some root pathogens and is a desirable goal of mycorrhizal management.

2.3.2 Stimulatory or inhibitory compounds

Soil organisms are known to produce an arsenal of biochemically active compounds like antibiotics, vitamins, and growth regulators, and these compounds can impact mycorrhizal function (Vancura 1986). A recent review (Gryndler 2000) illustrates the diversity of interactions between AM fungi and other soil organisms involving both stimulatory and inhibitory compounds. For example, an isolate of *Aspergillus niger* that produced substances similar to indole-3-acetic acid and gibberellic acid was shown to increase the fitness of both *Glomus fasciculatum* and the host plant

(Manjunath *et al*. 1981). In contrast, unidentified compounds produced by another isolate of *A. niger* were shown to inhibit spore germination and hyphal growth of *Glomus mosseae* (McAllister *et al*. 1995).

2.3.3 Modification of rhizosphere chemistry

Soil bacteria and fungi also impact mycorrhizal function by modifying soil chemistry. For example, synergistic relationships have been observed between AM fungi and P-solubilising bacteria (Barea *et al*. 1975, Piccini and Azcon 1987). These bacteria are thought to increase the solubility of calcium phosphate through acidification of the rhizosphere with organic acids (Kim *et al*. 1998). Enzymatic activities of rhizosphere organisms also generate synergistic relationships with AM fungi (Camprubi *et al*. 1995). For example, Tarafdar and Marschner (1995) found that extracellular phosphatase produced by *Aspergillus fumigans* increased P uptake and growth of wheat inoculated with *G. mosseae*.

2.3.4 Grazing extraradical hyphal networks

Microarthropod grazers may also be important mediators of mycorrhizal function. Most subterranean species of collembola feed heavily, if not exclusively on soil fungi. Some studies suggest that collembola could be important regulators of AM function because grazing on extraradical hyphal networks could seriously reduce the nutrient uptake capacity of AM fungi and potential benefit to plants (Warnock *et al*. 1982, Finlay 1985, Thimm and Larink 1995). However, other studies indicate that when given the choice, collembola avoid eating AM hyphae and much prefer to feed on nonmycorrhizal fungi (Klironomos *et al*. 1999). A recent review (Gange 2000) explores the complexity of collembola-AM fungal interactions and suggests that the grazing of collembola on nonmycorrhizal fungi may indirectly benefit AM fungi and host plants and stimulate nutrient cycling. This review also cautions against generalising too much from the current pool of literature because, to date, nearly all of the studies of AM-collembola interactions have used a single, easily cultureable collembola species: *Folsomia candida*. Many more studies need to be conducted that incorporate a wider diversity of collembola as well as other fungal grazers, such as fungivorous nematodes and mites, before the full impact of AM-grazer interactions on mycorrhizal function can be understood.

2.3.5 Modification of soil structure

Large, earthmoving soil organisms such as ants, earthworms, and gophers modify soil structure and impact propagule densities of AM fungi

(Allen 1991, Friese and Allen 1993). Mycorrhizal effects on soil structure also impact other soil organisms. As discussed previously, AM fungal hyphae facilitate the formation and stabilisation of soil aggregates. Andrade *et al.* (1998) used a split-pot design to show that soil populations of bacteria, actinomycetes and fungi all responded positively to the structural modifications caused by AM fungal hyphae. Total microbial populations were not correlated with AM root colonisation directly, but were strongly correlated with the increased aggregation caused by the AM fungi. Highly aggregated soil will be more aerobic and have a higher moisture holding capacity than soil with few stable aggregates. Both of these factors are likely to strongly influence soil microbial populations and indirectly feedback on mycorrhizal function.

2.3.6   Dispersal or destruction of propagules

Ants, grasshoppers, earthworms, millipedes, mites, and other soil animals are known to be important dispersal agents of AM fungi (Allen 1991). For example, Klironomos and Moutoglis (1999) showed that the collembola *Folsomia candida* increased the dispersal range of *Glomus etunicatum* by at least 30 cm. On the other hand, AM fungi can also be vectors for other soil organisms (Gryndler 2000). For example, Bianciotto *et al.* (1996) found that several strains of rhizobia and pseudomonads adhere to the surface of AM fungal spores, hyphae, and auxiliary cells and thus, AM fungi may transport these soil microbes throughout the soil.

From a negative perspective, interactions can also destroy propagules. Soil animals such as mites, collembola and worms can destroy AM fungal propagules through direct ingestion or piercing and sucking out the spore contents (Hetrick 1984). Also, bacteria, actinomycetes, and fungi are known to degrade spore walls and reduce their viability (Ross and Ruttencutter 1977, Ames *et al.* 1989).

2.4   Impacts of Arbuscular Mycorrhizas on Nutrient Cycling

Sustainable natural and managed systems efficiently recycle essential nutrients and minimise losses through erosion, leaching, or volatilisation. All of the functions of mycorrhizas (Table 1) may influence nutrient flux within ecosystems. Because AM fungi are often among the largest consumers of net primary production, they immobilise a tremendous quantity of nutrients, and the rate at which their tissues decompose will impact nutrient availability (Allen 1991). Also, intact networks of AM mycelia act as conduits for nutrient transfer within plant communities and may be important in reducing leaching losses (Read *et al.* 1985). The role of

AM fungi in direct acquisition of nutrients from organic matter is controversial (Hodge *et al.* 2000), but their indirect roles through plant nutrition and microbial communities can be substantial and need to be considered in ecosystem-level management.

3. INFLUENCE OF MANAGEMENT PRACTICES UPON FUNCTION OF AM FUNGI

3.1 Effects of Tillage and Soil Disturbance upon AM Fungi

Given that extraradical hyphae are both the inorganic nutrient absorbing organ of the mycorrhiza and an important component of the inoculum of AM fungi in the soil, soil disturbance can affect both the inoculum potential of the soil and the ability of the mycorrhiza to take up nutrients. In addition, tillage can affect the distribution of AM fungi through the soil profile.

First, severe soil disturbance, such as moldboard ploughing, can greatly affect the distribution of AM fungi within the plough layer of soil. Inoculum of AM fungi in undisturbed soil or at the end of a growing season in agricultural soil is found primarily in the top 8-15 cm of soil (Smith 1978, An *et al.* 1990, Abbott and Robson 1991). Moldboard ploughing would transport this inoculum to greater depths due to inversion of the soil (Smith 1978).

The rapid colonisation of a newly germinated seedling can depend to a large extent upon the intact network of extraradical mycelium already present in the soil. This hyphal network is built and destroyed with each tillage and planting cycle in an agricultural soil under conventional tillage. This affects the rate of colonisation of young seedlings by AM fungi. A common observation is greater colonisation of roots of seedlings in no-till soils early in the growing season relative to those in paired, tilled plots (Galvez *et al.* 1995, Kabir *et al.* 1997, McGonigle and Miller 1993) (Table 4), though this may not always happen (Miller *et al.* 1995). Two situations in which soil disturbance may not affect colonisation of roots by AM fungi are when the majority of the inoculum is in the form of spores, which remain viable after disturbance (Jasper *et al.* 1991) and when inoculum levels in the soil are low (McGonigle and Miller 2000). Characteristics of both the host crop and fungal symbionts should be considered in future studies of this phenomenon (McGonigle and Miller 2000). For example, genera of AM fungi differ in both the hyphal growth possible from a germinated spore and in the ability of infected root pieces or extraradical hyphae to act as inoculum (Biermann and Linderman 1983). The relative proportions of these groups within the AM fungus community would influence the response to tillage.

Soil disturbance or tillage more consistently affects the mycorrhiza-mediated P uptake of plants whether or not a concomitant decrease in colonisation occurred. This has been demonstrated in greenhouse (Evans and Miller 1990, Miller 2000) and field experiments (McGonigle *et al.* 1990, McGonigle and Miller 1996). There is a general consensus on the mechanism behind this phenomenon. The plant sown into the undisturbed, pre-existing network of extraradical mycelium becomes colonised and is then 'plugged into' an already extensive nutrient absorbing organ of the mycorrhiza. There may not need to be great levels of inoculum for this to occur. The plant in the undisturbed soil will exhibit increased P status early in the growing season (Table 4). However, the hyphal network in the disturbed soil eventually redevelops allowing these plants to 'catch up' as early as the 6-leaf stage in maize (McGonigle and Miller 1993). Indeed, the early season enhancement in P uptake for no-till *vs* conventionally tilled maize does not translate into increased growth and yield (Miller *et al.* 1995), possibly due to reduced soil temperatures in no-tilled soils (Miller 2000). Also, tillage is likely to select for different AM fungal species (Johnson unpublished observation, Jansa *et al.* 2001), with different symbiotic function. More research is needed to increase the yield of no-till crops so as to encourage this management practice with a wide range of environmental benefits.

Table 4 Effect of moldboard plough (MP) and no-till (NT) upon maize shoot P concentration and colonisation of roots by AM fungi in the field.*

Days after Planting	Shoot P (mg kg^{-1})		Root length with arbuscules (%)	
	MP	NT	MP	NT
25	0.642 b	0.752 a	11 b	27 a
32	0.344 b	0.480 a	27 b	45 a
48	0.442 a	0.441 a	44 a	55 a

*Numbers in a row, for a given pairwise comparison, followed by the same number are not significantly different (p=0.05). Adapted from McGonigle and Miller 1993.

Tillage should also be expected to interact with the soil aggregation function of AM fungi. A three year transition from tillage to no-till was studied in a silt loam soil (Wright *et al.* 1999). Both soil aggregate stability and glomalin levels in the soil were greater for no-till than tillage treatments, and the effect was greater with successive years of no-till (Table 5).

Table 5 Aggregate stability (0-5 cm depth) and glomalin content for a silt loam soil in transition from tillage to no-till.*

Treatment	Aggregate stability (%)	Total glomalin (mg g^{-1} aggregates)
No-till 3 yrs	37.7 a	1.567 a
No-till 2 yrs	30.8 ab	1.389 ab
No-till 1 yr	25.0 b	1.323 bc
Plough tillage	16.7 c	1.195 c

*Numbers in the same column followed by the same letter are not significantly different (p=0.05). Adapted from Wright *et al.* 1999.

3.2   Effects of Crop Management upon AM Fungi

The community of AM fungi in agricultural soil is also influenced by the choice of crop host and crop rotation history. In addition, the presence and length of fallow periods, or the presence of over-wintering or fallow cover crops, have significant effects on the composition of AM fungal communities, which in turn affects the productivity of the soil.

Greenhouse studies showed that AM fungi proliferate more in the presence of one host than another, and that preferred hosts differ among AM fungal species (Hetrick and Bloom 1986). This also occurs in the field where the abundance of spores of certain AM fungal species will rise and fall according to the cycle of the crop rotation (Hayman *et al.* 1975, An *et al.* 1993, Hendrix *et al.* 1995). For example, *Gigaspora gigantea* spores were more numerous in the autumn following maize (3.5 spores 50 cm^{-3}) than following small grains or a vegetable crop (0.5 spores 50 cm^{-3}) (Douds *et al.* 1997). However, this species was more prevalent following soybean at another site (An *et al.* 1993), underscoring the important interaction with soil characteristics, which has been noted elsewhere (Johnson *et al.* 1991).

The species composition of the AM fungal community can have important ramifications for the biological fertility of soils. Certain species, notably those from the genus *Gigaspora*, are more often associated with well aggregated soils (Miller and Jastrow 1992), and therefore may play a stronger role in stabilising macroaggregates than other genera. As abundance of AM fungal species changes within the rotation, so may the susceptibility of the soil to erosion. These conclusions are based upon spore populations. Researchers have not had the proper tools to allow them to consider the relative contribution of the high or low abundance sporulators to the length of extraradical hyphae in the soil (for discussion, see Douds and Millner 1999). There have been no field-based descriptions of AM fungal communities based upon extraradical hyphal networks, i.e. the structures that actually do the work of nutrient uptake and soil stabilisation. Some researchers have successfully discriminated among selected AM fungi based upon morphology of intraradical structures (Abbott 1982), and this was useful in greenhouse studies (Abbott and Robson 1984). Merryweather and Fitter (1998) described the arbuscular mycorrhizas of *Hyacinthoides non-scripta* and assigned genus and sometimes species designations based upon intraradical morphology, but later found discrepancies between morphology-based identifications and those based upon molecular techniques (Helgason *et al.* 1998-a). If this was due to phenotypic plasticity of the AM fungi, then molecular techniques hold the greater promise for future study. Although there have been recent successes in greenhouse pot experiments (Jacquot *et al.* 2000, Kjoller and Rosendahl 2000pa), there is need for field application of molecular techniques for the identification and quantification of the extraradical phase of mycorrhizas.

Another aspect of 'functional biodiversity' of AM fungi to be considered with changes in the community is in the exploration of the soil for nutrients. Smith *et al.* (2000) and the related commentary by Koide (2000) recently demonstrated this. *Scutellospora calospora* enhanced P uptake by *Medicago truncatula* from soil close to the root while *Glomus caledonium* enabled access to more distant soil. The whole subject of functional diversity of AM fungi requires further research. There are approximately 160 described AM fungus species, upwards of 26 at a site (Ellis *et al.* 1992), and beyond the work noted above, the prevailing view is that they more or less occupy the same niche in the soil (Dodd *et al.* 2000).

There is evidence that the efficacy of the entire community of AM fungi can change due to cropping sequence and that crop monocultures may generate fungal communities that do not enhance crop performance (Schenck *et al.* 1989). This phenomenon may partly explain the reduction of yield that occurs with continuous monocultures of maize, soybean (Johnson *et al.* 1992), and tobacco (Hendrix *et al.*1992). Feldmann *et al.* (1991) also reported that AM fungi from a monoculture of *Hevea* spp were ineffective at growth promotion of both *Hevea* seedlings and maize compared to those from a nearby natural area.

The presence of weed host plants in an agricultural field can serve to offset the negative effect of a monoculture upon the AM fungal community. Populations of spores of AM fungi were positively correlated to the presence of weeds in lettuce fields (Miller and Jackson 1998). Feldmann and Boyle (1999) found beneficial effects of weeds, not only upon diversity of AM fungi in the soil, but upon the efficacy in enhancing plant growth, overcompensating for any yield reduction of maize due to competition with the weeds. Other studies have noted no increased spore populations or diversity with increasing weed levels (Kurle and Pfleger 1996).

One of the crop management practices most harmful to AM fungi is a nonhost plant, such as *Brassica* or *Lupinus* in the crop rotation (Blaszkowski 1995, Harinikumar and Bagyaraj 1988). These crops resist colonisation by AM fungi, thereby restricting the carbon flow to these obligate symbionts. This results in lower levels of inoculum and less colonisation of the next mycotrophic crop in the rotation (Gavito and Miller 1998). The growth of some high value vegetable crops, with wide spacings and strict weed control, can also depress AM fungal populations due to restriction of available root biomass for the symbiosis. Less inoculum of AM fungi was present in soil following a crop of *Capsicum annuum* than following small grains or maize (Douds *et al.* 1997).

Long fallow periods without plant cover are detrimental to AM fungi. A striking example of this is long-fallow disorder in Australia (Thompson 1987, 1991). Long fallows of 11-14 months may be necessary in semi-arid climates to allow for replenishment of soil moisture for the next crop. A wide range of crop species, among which *Linum usitatissimum* is one of the most sensitive, can grow poorly after long fallows and exhibit P and Zn deficiency. Poor crop growth was correlated to reduced root colonisation by AM fungi due to the reduction in inoculum caused by long fallow in northern Australia (Thompson 1987), but this effect was not observed in southern Australia (Ryan and Angus 2001).

On the other hand, one of the most AM fungus-friendly crop management practices, besides reduced tillage, is inclusion of over wintering cover crops in a crop rotation. One of the primary differences between 'low-input, sustainable' agriculture and 'chemical-based, conventional' agriculture is that the proportion of the year with live plant cover is much greater in the low-input than conventional system (Douds *et al.* 1993). Soils of the low-input farming system studied at the Rodale Institute Experimental Farm are covered with live plants approximately 70% of an average year *vs* 40% for a conventional maize-soybean rotation. This is primarily due to over wintering cover crops. Spores in bare fallowed soils may be induced to germinate during mild late fall or early spring periods, after crop senescence and before the planting of the next crop. Several such germinations cause the spore to drain its carbohydrate and lipid reserves, limiting its ability to colonise roots of the next crop when they are finally available. The cover crop, in addition to retarding soil erosion and replenishing and retaining soil N, serves as an interim host plant for the AM fungi. This results in greater amounts of infective inoculum for the next cash crop (Galvez *et al.* 1995; Boswell *et al.* 1998).

The use of transgenic plants as a crop management technique to control insect pests is practiced widely in some parts of the world, particularly in the US (Stewart *et al.* 2000). Future application of this technology to modify the human nutritional value of crops is likely (Ye *et al.* 2000). Oger *et al.*

(2000) showed that *Lotus corniculatus* with transgenes for opine production were capable of altering the rhizosphere microflora. The future likelihood of the development of transgenic plants resistant to root diseases, encouraged due to future restrictions on the use of chemicals such as methyl bromide, underscores the need for studies of the effects of transgenic plants upon beneficial soil microbes such as AM fungi (Glandorf *et al.* 1997).

3.3   Nutrient Management

Mycorrhizal fungi are generally very sensitive to phosphate enrichment. Their responses to fertilisation are mediated by their host plants and are strongly influenced by edaphic properties and chemical composition of the fertiliser.   Fertilisation usually decreases AM fungus colonisation in agricultural soils but in extremely nutrient deficient soils, it sometimes increases colonisation (Hayman 1975). These contrasting responses suggest that plant nutrition mediates mycorrhizal responses to fertiliser. Severely nutrient deficient plants are stunted and can supply little photosynthate to AM fungi. Fertilisation of these systems will increase plant size and their ability to provide AM fungi with carbon compounds. Plants usually preferentially allocate photosynthate to the organs that maximise acquisition of the resources that are most limiting to growth (Chapin 1980, Tilman 1988). Thus, when plants are strongly limited by soil resources, relatively more photosynthate is allocated to their roots; and, when they are more strongly limited by light, relatively more is allocated to shoots. Fertilisation reduces limitation by soil nutrients and induces plants to allocate less carbon to roots, root exudates and AM fungi. Most agricultural soils have moderate to high nutrient contents and this is why fertilisation usually decreases AM colonisation in agricultural systems.   Together, the chemistry of soils and fertiliser supplements will control the nutrient status of plant hosts and they will ultimately control mycorrhizal responses to fertilisation.

Studies show interactions in mycorrhizal responses to P, N, and potassium (K) enrichment, indicating that the relative availability of these essential nutrients is important to mycorrhizal function (Saif 1986, Gryndler *et al.* 1990). Although AM fungi are best known for their uptake of P and immobile micronutrients, mycorrhizal uptake of N, particularly as NH_4, is also well documented (Ames *et al.* 1983). Enrichment of P does not necessarily reduce AM fungal colonisation in N-limited plants, but it does reduce colonisation when N levels are adequate. This suggests that P:N ratio is an important factor governing AM responses to nutrient enrichment (Hepper 1983, Sylvia and Neal 1990).

In general, conventional agricultural systems enriched with high inputs of inorganic fertilisers have lower AM fungus activity than organically managed systems enriched with low inputs of farmyard manure or legume cover crops (Sattelmacher *et al.* 1991, Douds *et al.* 1993, Galvez *et al.* 1995). A study of 24 fields at 13 farms showed that fields enriched with inorganic fertilisers had significantly lower AM fungal colonisation and fewer spores than fields managed with no inputs or enriched with manure and slurry (Eason *et al.* 1999). This study took the important next step of assessing the symbiotic functioning of the AM associations in these fields. Spores were sieved from the 24 soils and approximately 1,000 spores from each soil were used to inoculate *Allium ameloprasum* and *Trifolium repens* grown in irradiated soil in a greenhouse. On average, these crops had significantly larger shoots when inoculated with spores from organic fields than from conventional fields (Table 6). These findings suggest that the AM fungi in the low-fertility organically managed soils were more beneficial to their hosts than those in the high-fertility conventionally managed soils. Analyses of spore populations in other systems indicate that fertilisation changes the species composition of AM fungal communities (Johnson 1993, Egerton-Warburton and Allen 2000). These results provide support for the hypothesis that fertilisation selects for less mutualistic AM fungi (Johnson 1993). Theoretically, nutritional mutualisms would be expected to be selected in nutrient deficient systems and fertilisation would be expected to eliminate the benefits conferred by such a relationship and set the stage for more parasitic interactions (Johnson *et al.* 1997, Hoekesema and Bruna 2000).

More studies are needed to link the composition of AM fungal communities with their symbiotic function. Because the benefits that plants gain from mycorrhizas are often unrelated to root colonisation and spore densities (McGonigle 1988), future field-based research needs to systematically examine fertiliser impacts on mycorrhizal functioning, across a range of crops and soil types, to provide the information that is necessary to effectively coordinate management of mycorrhizas and fertilisers in a sustainable manner.

Table 6 Responses of *Allium* and *Trifolium* to inoculation with AM fungal spores from grassland soil under conventional management (CM) or organic management (OM)*.

	Allium ameloprasum		*Trifolium repens*	
	CM	OM	CM	OM
Total shoot weight (mg)	0.39b	0.55a	9.44b	10.30a
AM infection (% root length)	56.9a	64.0a	54.5a	63.3a

*Numbers in a row, for a given pairwise comparison, followed by the same number are not significantly different (α=0.05). Adapted from Eason *et al.* 1999.

3.4   Effects of Synthetic Pesticides upon AM Fungi

Chemical pesticides applied to agricultural soils throughout the production cycle may have variable affects on AM function. Soils may be fumigated prior to planting. Most of these fumigants, including dazomet (Mark and Cassells 1999) and methyl bromide / chloropicrin (McGraw and Hendrix 1984) are also effective at killing indigenous AM fungi. Although enhanced growth of the following crop due to control of pathogens is the expected result of fumigation, stunted growth and P, Cu, or Zn deficiency may also occur. This has been linked to the destruction of AM fungi because inoculation with AM fungi relieves the stunting (see thorough review by Menge 1982).

Seeds sown into agronomic soils may also be coated with fungicides. These fungicides were shown to have no effect on the development of mycorrhizas on the seedlings (Spokes *et al.* 1989). Fungicides may also be applied to soils prior to or during plant growth. These affect AM fungi to varying degrees, and species of AM fungi differ in their susceptibility (Schreiner and Bethlenfalvay 1997). Further, the extraradical hyphae, i.e. the nutrient absorbing organ of the mycorrhiza, appears to be the most susceptible to fungicide application (Kjoller and Rosendahl 2000b, Larsen *et al.* 1996).

4.   CONCLUSION

Mycorrhizal effects on plant production are mediated by complex interactions among soil properties, plant genotypes, AM fungal genotypes, and the physical and biotic environment. The result of these interactions over time is the selection of communities of soil organisms that may or may not maximise crop production. Thus, in the context of developing management strategies to maximise AM benefits, it is necessary to analyse these associations from an evolutionary perspective and consider them dynamic systems integrating interactions at molecular, population, community and ecosystem scales (Miller and Kling 2000).

5.   ACKNOWLEDGEMENTS

This manuscript was prepared in part with support from grant DEB 98-06529 from the National Science Foundation to JCN.

6. REFERENCES

Abbott L K 1982 Comparative anatomy of vesicular-arbuscular mycorrhizas formed on subterranean clover. Australian Journal of Botany 30: 485-499.

Abbott L K and Gazey C 1994 An ecological view of the formation of VA mycorrhizas. Plant and Soil 159: 69-78.

Abbott L K and Robson A D 1982 Infectivity of vesicular-arbuscular mycorrhizal fungi in agricultural soils. Australian Journal of Agricultural Research 33: 1049-1059.

Abbott L K and Robson A D 1984 Colonisation of the root system of subterranean clover by three species of vesicular-arbuscular mycorrhizal fungi. New Phytologist 96: 275-281.

Abbott L K and Robson A D 1985 Formation of external hyphae in soil by four species of vesicular-arbuscular mycorrhizal fungi. New Phytologist 99: 245-255.

Abbott L K and Robson A D 1991 Factors influencing the occurrence of vesicular-arbuscular mycorrhizas. Agricriculture Ecosystems and Environment 35: 121-150.

Afek U, Menge J A and Johnson E L V 1990 Effect of *Pythium ultimum* and metalaxyl treatments on root length and mycorrhizal colonisation of cotton, onion, and pepper. Plant Disease 74: 117-120.

Allen M F 1982 Influence of vesicular-arbuscular mycorrhizae on water movement through *Bouteloua gracilis* (H.B.K.). Lag ex Steud. New Phytologist 91: 191-196.

Allen M F 1991 The ecology of mycorrhizae. Cambridge University Press. New York.

Allen M F, Smith W K, Moore Jr T S and Christensen M 1981 Comparative water relations and photosynthesis of mycorrhizal and non-mycorrhizal *Bouteloua gracilis* H.B.C. New Phytologist 88: 683-693.

Ames R N, Reid C P P, Porte L K and Cambardella C 1983 Hyphal uptake and transport of nitrogen from two ^{15}N-labelled sources by *Glomus mosseae*, a vesicular-arbuscular mycorrhizal fungus. New Phytologist 95: 381-396.

Ames R N, Reid C P P and Ingham E R 1984 Rhizosphere bacterial populations responses to root colonization by a vesicular-arbuscular mycorrhizal fungus. New Phytologist 96: 555-563.

Ames R N, Mihara K L and Bayne H G 1989 Chitin-decomposing actinomycetes associated with a vesicular-arbuscular mycorrhizal fungus from a calcareous soil. New Phytologist 111: 67-71.

An Z-Q, Grove J H, Hendrix J W, Hershman D E and Henson G T 1990 Vertical distribution of endogonaceous mycorrhizal fungi associated with soybean as affected by soil fumigation. Soil Biology and Biochemistry 22: 715-719.

An Z-Q, Hendrix J W, Hershman D E, Ferriss R S and Henson G T 1993 The influence of crop rotation and sol fumigation on a mycorrhizal fungal community associated with soybean. Mycorrhiza 3: 171-182.

Andrade G, Mihara K L, Linderman R G and Bethlenfalvay G J 1997 Bacteria from rhizosphere and hyphosphere soils of different arbuscular mycorrhizal fungi. Plant and Soil 192: 71-79.

Andrade G, Mihara K L, Linderman R G and Bethlenfalvay G J 1998 Soil aggregation status and rhizobacteria in the mycorrhizosphere. Plant and Soil 202: 89-96.

Augé R M 2000 Stomatal behaviour of arbuscular mycorrhizal plants. *In:* Arbuscular Mycorrhizas: Physiology and Function. Y Kapulnik and D D Douds Jr (eds.) pp. 201-237. Kluwer Academic Publishers. Dordrecht, The Netherlands.

Augé R M, Schekel K A and Wample R L 1986 Greater leaf conductance of well-watered VA mycorrhizal rose plants is not related to phosphorus nutrition. New Phytologist 103: 107-116.

Augé R M and Duan X 1991 Mycorrhizal fungi and nonhydraulic root signals of soil drying. Plant Physiology 97: 821-824.

Azcon-Aguilar C, Azcon R and Barea J M 1979 Endomycorrhizal fungi and *Rhizobium* as biological fertilizers for *Medicago sitiva* in normal cultivation. Nature 279: 325-327.

Bago G, Shachar-Hill Y and Pfeffer P E 2000 Dissecting carbon pathways in arbuscular mycorrhizas with NMR spectroscopy. *In:* Current Advances in Mycorrhizae Research. G K Podila and D D Douds Jr (eds.) pp. 111-126. American Society of Phytopathology Press. St. Paul, MN, USA.

Barea J M, Azcon R and Hayman D S 1975 Possible synergistic interactions between Endogone and phosphate solubilizing bacteria in low phosphate soils. *In:* Endomycorrhizas. F E Sanders, B Mosse and P B Tinker (eds.) pp. 409-418. Academic Press. London.

Barea J M, Azcon R and Azcon-Aguilar C 1992 Vesicular-arbuscular mycorrhizal fungi in nitrogen-fixing systems. Methods in Microbiology 24: 391-416.

Baylis G T S 1975 The magnolioid mycorrhiza and mycotrophy in root systems derived from it. *In:* Endomycorrhizas. F E Sanders, B Mosse and P B Tinker (eds.) pp. 373-389. Academic Press. London.

Benhamou N, Fortin J A, Hamel C, St-Arnaud M and Shatilla A 1994 Resistance responses of mycorrhizal Ri T-DNA transformed carrot roots to infection by *Fusarium oxysporum* F. sp. *chrysanthemi*. Phytopathology 84: 958-968.

Bethlenfalvay G J, Franson R L, Brown M S and Mihara K L 1989 The *Glycine-Glomus-Bradyrhizobium* symbiosis. IX. Nutritional, morphological and physiological responses of nodulated soybean to geographic isolates of the mycorrhizal fungus *Glomus mosseae*. Physiologia Plantarum 76: 226-232.

Bethlenfalvay G J, Schüepp H 1994 Arbuscular mycorrhizas and agrosystem stability. *In:* Impact of Arbuscular Mycorrhizas on Sustainable Agriculture and Natural Ecosystems. S Gianinazzi and H Schüepp (eds.) pp. 117-131. Birkhauser Verlag. Basel, Switzerland.

Bethlenfalvay G J and Linderman R G 1992 Mycorrhizae in sustainable agriculture. ASA Special Publication No. 54. American Society of Agronomy. Madison Wisconsin, USA.

Bethlenfalvay G J, Barea J-M 1994 Mycorhizae in sustainable agriculture. I. Effects on seed yield and soil aggregation. American Journal of Alternative Agriculture 9: 157-160.

Bianciotto V, Minerdi D, Perotto S and Bonfante P 1996 Cellular interactions between arbuscular mycorrhizal fungi and rhizosphere bacteria. Protoplasma. 193: 123-131.

Biermann B and Linderman R G 1983 Mycorrhizal roots, intraradical vesicles and extra-radical vesicles as inoculum. New Phytologist 95: 97-105.

Biro B, Koves-Pechy K, Voros I, Takacs T, Eggenberger P and Strasser R J 2000 Interrelations between *Azospirillum* and *Rhizobium* nitrogen-fixers and arbuscular mycorrhizal fungi in the rhizosphere of alfalfa in sterile, AMF-free or normal soil conditions. Applied Soil Ecology 15: 159-168.

Blaszkowski J 1995 The influence of pre-crop plants on the occurrence of arbuscular mycorrhizal fungi (Glomales) and *Phialophora graminicola* associated with roots of winter X *Triticosecale*. Acta Mycologia 30: 213-222.

Blee K A and Anderson A J 1998 Regulation of arbuscule formation by carbon in the plant. Plant Journal 16: 523-530.

Bolan N S 1991 A critical review on the role of mycorrhizal fungi in the uptake of phosphorus by plants. Plant and Soil. 134: 189-207.

Boswell E P, Koide R T, Shumway D L and Addy H D 1998 Winter wheat cover cropping, VA mycorrhizal fungi and maize growth and yield. Agriculture Ecosystems and Environment 67: 55-65.

Bryla D R and Duniway J M 1998 The influence of the mycorrhizal fungus *Glomus etunicatum* on drought acclimation in safflower and wheat. Physiology of Plants 104: 87-96.

Camprubi A, Calvet C and Estaun V 1995 Growth enhancement of *Citrus reshni* after inoculation with *Glomus intraradices* and *Trichoderma aureoviride* and associated effects on microbial populations and enzyme activity in potting mixes. Plant and Soil 173: 233-238.

Caron M, Fortin J A and Richard C 1986 Effect of phosphorus concentration and *Glomus intraradices* on Fusarium crown and root rot of tomatoes. Phytopathology 76: 942-946.

Chapin F S 1980 The mineral nutrition of wild plants. Annual Review of Ecological Systems 11: 233-260.

Christenson H and Jakobsen I 1993 Reduction of bacterial growth by a vesicular-arbuscular mycorrhizal fungus in the rhizosphere of cucumber (*Cucumis sativus L.*). Biology and Fertility of Soils 15: 253-258.

Cordier C, Gianinazzi S and Gianinazzi-Pearson V 1996 Colonization patterns of root tissues by *Phytophthora nicotianae* var *parasitica* related to reduced disease in mycorrhizal tomato. Plant and Soil. 185: 223-232.

Dar G H, Zargar M Y and Beigh G M 1997 Biocontrol of Fusarium root rot in the common bean *Phaselous vulgaris* by using symbiotic *Glomus mosseae* and *Rhizobium leguminosarum*. Microbial Ecology 34: 74-80.

Degens B P, Sparling G P and Abbott L K 1994 The contribution from hyphae, roots and organic carbon constituents to the aggregation of a sandy loam under long-term clover-based and grass pastures. European Journal of Soil Science 45: 459-468.

Dodd J C, Boddington C L, Rodriguez A, Gonzalez-Chavez C and Mansur I 2000 Mycelium of arbuscular mycorrhizal fungi (AMF) from different genera: form, function and detection. Plant and Soil. 226: 131-151.

Douds D D, Janke R R and Peters S E 1993 VAM fungus spore populations and colonization of roots of maize and soybean under conventional and low-input sustainable agriculture. Agriculture, Ecosystems and Environment 43: 325-335.

Douds D D, Galvez L, Franke-Snyder M, Reider C and Drinkwater L E 1997 Effect of compost addition and crop rotation upon VAM fungi. Agriculture, Ecosystems and Environment 65: 257-266.

Douds D D Jr and Millner P D 1999 Biodiversity of arbuscular mycorrhizal fungi in agroecosystems. Agriculture, Ecosystems and Environment 74: 77-93.

Douds D D Jr, Pfeffer P E and Shachar-Hill Y 2000 Carbon partitioning, cost, and metabolism of arbuscular mycorrhizas. *In* Arbuscular Mycorrhizas Physiology and Function. Y Kapulnik and D D Douds Jr (eds.) pp. 107-129. Kluwer Academic Publishers. Dordrecht, The Netherlands.

Eason W R, Scullion J and Scott E P 1999 Soil parameters and plant responses associated with arbuscular mycorrhizas from contrasting grassland management regimes. Agriculture, Ecosystems and Environment 73: 245-255.

Ebel R C, Wellbaum G E, Gunatilaka M, Nelson T and Augé R M 1996 Arbuscular mycorrhizal symbiosis and nonhydraulic signalling of soil drying in *Vigna unguiculata* (L.) Walp. Mycorrhiza. 6: 119-127.

Egerton-Warburton L M and Allen E B 2000 Shifts in arbuscular mycorrhizal communities along an anthropomorphic nitrogen deposition gradient. Ecol. Appl. 10: 484-496.

Ellis J R 1998 Post flood syndrome and vesicular-arbuscular mycorrhizal fungi. Journal of Production Agriculture 11: 200-204.

Ellis J R, Roder W and Mann S C 1992 Grain sorghum - soybean rotation and fertilization influence on vesicular-arbuscular mycorrhizal fungi. Soil Science Society of America Journal 56: 783-794.

Evans D G and Miller M H 1990 The role of the external mycelial network in the effect of soil disturbance upon vesicular-arbuscular mycorrhizal colonization of maize. New Phytologist 114: 65-72.

Feldmann F, Werlitz J, Junqueira N T V and Leiberei R 1991 Mycorrhizal populations of monocultures are less effective to the crop than those of natural stands! Third European Symposium on Mycorrhizas. August 19-23, 1991. Sheffield, UK.

Feldmann F and Boyle C 1999 Weed-mediated stability of arbuscular mycorrhizal effectiveness in maize monocultures. Journal of Applied Botany - Angew. Bot. 73: 1-5.

Filion M, St-Arnaud M and Fortin J A 1999 Direct interaction between the arbuscular mycorrhizal fungus *Glomus intraradices* and different rhizosphere microorganisms. New Phytologist. 141: 525-533.

Finlay R D 1985 Interactions between soil microarthropods and endomycorrhizal associations of higher plants. *In:* Ecological Interactions in Soil. A H Fitter, D Atkinson, D J Read and M Busher (eds.) pp. 319-331. Blackwell Scientific Publications. London.

Fitter A H 1991 Costs and benefits of mycorrhizas: Implications for functioning under natural conditions. Experientia 47: 350-355.

Friese C F and Allen M F 1993 The interaction of harvester ants and vesicular-arbuscular mycorrhizal fungi in a patchy semi-arid environment: the effects of mound structure on fungal dispersion and establishment. Functional Ecology 7: 13-20.

Galvez L, Douds D D, Wagoner P, Longnecker L R, Drinkwater L E and Janke R R 1995 An overwintering cover crop increases inoculum of VAM fungi in agricultural soil. American Journal of Alternetive Agriculture 10: 152-156.

Gange A 2000 Arbuscular mycorrhizal fungi, collembola and plant growth. Trends in Ecological Evolution 15: 369-372.

Garbaye J 1994 Helper bacteria: a new dimension to the mycorrhizal symbiosis. Tansley Review No. 76. New Phytologist 128: 197-210.

Garland J L and Mills A L 1991 Classification and characterization of heterotrophic microbial communities on the basis of patterns of community-level sole-carbon source utilization. Applied Environmental Microbiology 57: 2351-2359.

Gavito M E and Miller M H 1998 Changes in mycorrhiza development in maize induced by crop management practices. Plant and Soil. 198: 185-192.

George E, Haeussler K-U, Vetterlein D, Gorgus E and Marschner H 1992 Water and nutrient translocation by hyphae of *Glomus mosseae*. Canandian Journal of Botany 70: 2130-2137.

Gianinazzi S a nd Schüepp H 1994 Impact of arbuscular mycorrhizas on sustainable agriculture and natural ecosystems. Birkhauser Verlag. Basel, Switzerland.

Giovannetti M, Tosi L, Della Torre G and Zaggerini A 1991 Histological, physiological and biochemical interactions between vesicular-arbuscular mycorrhizae and *Thielaviopsis brasicola* in tobacco plants. Journal of Phytopathology 131: 265-274.

Glandorf D C M, Bakker P A H M and van Loon L C 1997 Influence of the production of antibacterial and antifungal proteins by transgenic plants on the saprophytic soil microflora. Acta Bot. Neerl. 46: 85-104.

Graham J H 2000 Assessing costs of arbuscular mycorrhizal symbiosis in agroecosystems. *In:* Current Advances in Mycorrhizae Research. G K Podila and D D Douds Jr (eds.) pp. 127-140. Amer. Soc. Phytopathol. Press. St. Paul, MN.

Graham J H and Egel D S 1988 Phytophthora root rot development on mycorrhizal and phosphorus fertilized nonmycorrhizal sweet orange seedlings. Plant Disease 72: 611-614.

Graham J H and Abbott L K 2000 Wheat responses to aggressive and non-aggressive arbuscular mycorrhizal fungi. Plant and Soil 220: 207-218.

Gryndler M 2000 Interactions of arbuscular mycorrhizal fungi with other soil organisms. *In:* Arbuscular Mycorrhizas: Physiology and Function. Y Kapulnik and D D Douds Jr (eds.) pp. 239-262. Kluwer Academic Publishers. Dordrecht, The Netherlands.

Gryndler M, Lestina J, Moravec V, Prikyl Z and Lipavsky J 1990 Colonization of maize roots by VAM-fungi under conditions of long-term fertilization of varying intensity. Agricriculture, Ecosystems and Environment 29: 18-186.

Harinikumar K M and Bagyaraj D J 1988 Effect of crop rotation on native vesicular-arbuscular mycorrhizal propagules in soil. Plant and Soil. 110: 77-80.

Hartnett D C and Wilson G W T 1999 Mycorrhizae influence plant community structure and diversity in tallgrass prairie. Ecology 80: 1187-1195.

Hayman D S 1975 The occurrence of mycorrhiza in crops as affected by soil fertility. *In:* Endomycorrhizas. F E Sanders, B Mosse and P B Tinker (eds.) pp. 495-509. Academic Press. London.

Hayman D S and Mosse B 1972 Plant growth responses of vesicular-arbuscular mycorrhiza. III. Increased plant uptake of labile P from soil. New Phytologist 71: 41-47.

Hayman D S, Johnson A M and Ruddlesdin I 1975 The influence of phosphate and crop species on *Endogone* spores and vesicular-arbuscular mycorrhiza under field conditions. Plant and Soil. 43: 489-495.

Helgason T, Merryweather J, Fitter A and Young P 1998 (a) Host preference and community structure of arbuscular mycorrhizal (AM) fungi in a semi-natural woodland. Proceedings of the Second International conference on Mycorrhiza, July 5-10, 1998. p. 81. Uppsala, Sweden.

Helgason T, Daniell T J, Husband R, Fitter A H and Young J P W 1998 (b) Ploughing up the wood-wide web? Nature 394: 431.

Hendrix J W, Guo B Z and An Z Q 1995 Divergence of mycorrhizal fungal communities in crop production systems. Plant and Soil. 170: 131-140.

Hendrix J W, Jones K J and Nesmith W C 1992 Control of pathogenic mycorrhizal fungi in maintenance of soil productivity by crop rotation. Journal of Production Agriculture 5: 383-386.

Hepper C M 1983 The effect of nitrate and phosphate on the vesicular-arbusclar mycorrhizal infection of lettuce. New Phytologist 93: 389-399.

Hetrick B A D 1984 Ecology of VA mycorrhizal fungi. *In:* VA Mycorrhiza. C L Powell and D J Bagyaraj (eds.) pp. 35-55. CRC Press. Boca Raton Florida, USA.

Hetrick B A D and Bloom J 1986 The influence of host plant on production and colonization ability of vesicular-arbuscular mycorrhizal spores. Mycologia. 78: 32-36.

Hetrick B A D, Wilson G W T and Todd T C 1992 Relationships of mycorrhizal symbiosis, rooting strategy, and phenology among tallgrass prairie forbs. Canadian Journal of Botany 70: 1521-1528.

Hetrick B A D, Wilson G W T and Cox T S 1993 Mycorrhizal dependence of modern wheat cultivars and ancestors: a synthesis. Canadian Journal of Botany 71: 512-518.

Hetrick B A D, Wilson G W T and Todd T C 1996 Mycorrhizal response in wheat cultivars: relationship to phosphorus. Canadian Journal of Botany 74: 19-25.

Hijri M, Hosny M, van Tuinen D and Dulieu H 1999 Intraspecific ITS polymorphism in *Scutellospora castanea* (Glomales, Zygomycotina) is structured within multinucleate spores. Fungal Genet. Biol. 26: 141-151.

Hodge A 2000 Microbial ecology of the arbuscular mycorrhiza. FEMS Microbiology Ecology 32: 91-96.

Hodge A, Robinson D and Fitter A 2000 Are microorganisms more effective than plants at competing for nitrogen? Trends in Plant Science 5: 304-308.

Hoeksema J D and Bruna E M 2000 Pursuing the big questions about interspecific mutualism: a review of theoretical approaches. Oecologia 125: 321-330.

Hosny M, Hijri M, Passerieux E and Dulieu H 1999 rDNA units are highly polymorphic in *Scutellospora castanea* (Glomales, Zygomycetes). Gene 226: 61-71.

INVAM 2001 International culture collection of arbuscular and vesicular arbuscular mycorrhizal fungi. Online: HTTP://INVAM.CAF.WVU.edu/

Jacquot E, van Tuinen D, Gianinazzi S and Gianinazzi-Pearson V 2000 Monitoring species of arbuscular mycorrhizal fungi in plants and in soil by nested PCR: application to the study of the impact of sewage sludge. Plant and Soil. 226: 179-188.

Jakobsen I 1994 Research approaches to study the functioning of vesicular-arbuscular mycorrhizas in the field. Plant and Soil 159: 141-147.

Jakobsen I and Rosendahl L 1990 Carbon flow into soil and external hyphae from roots of mycorrhizal cucumber plants. New Phytologist 115: 77-83.

Jansa J, Mozafar A, Ruh R, Anken T, Kuhn G, Sanders I and Frossard E 2001 Changes in community structure of AM fungi due to reduced tillage. Proceeding of the Third International Conference on Mycorrhizas. July 8-13, 2001. pp. C1 06 Adelaide, South Australia.

Jarvis A J and Davies W J 1998 The coupled response of stomatal conductance to photosynthesis and transpiration. Journal of Experimental Botany 49: 399-406.

Jasper D A, Abbott L K and Robson A D 1991 The effect of soil disturbance on vesicular-arbuscular mycorrhizal fungi in soils from different vegetation types. New Phytologist 118: 471-476.

Jastrow J D, Miller R M and Lussenhop J 1998 Contributions of interacting biological mechanisms to soil aggregate stabilization in restored prairie. Soil Biology and Biochemistry 30: 905-917.

Johnson N C 1993 Can fertilization select less mutualistic mycorrhizae? Ecol. Appl. 3: 749-757.

Johnson N C, Pfleger F L, Crookston R K and Simmons S R 1991 Vesicular-arbuscular mycorrhizas respond to corn and soybean cropping history. New Phytologist. 117: 657-664.

Johnson N C, Copeland P J, Crookston R K and Pfleger F L 1992 Mycorrhizae: possible explanation for yield decline with continuous corn and soybean. Agronomy Journal 84: 387-390.

Johnson N C, Graham J H and Smith F A 1997 Functioning and mycorrhizal associations along the mutualism-parasitism continuum. New Phytologist 135: 575-586.

Joner E J and Johansen A 2000 Phosphatase activity of external hyphae of two arbuscular mycorrhizal fungi. Mycological Research 104: 81-86.

Joner E J, Ravnskov S and Jakobsen I 2000 Arbuscular mycorrhizal phosphate transport under monoaxenic conditions. Biotechnology Letters 22: 1705-1708.

Jungk A and Claassen N 1986 Availability of phosphate and potassium as the result of interactions between root and soil in the rhizosphere. Zeits. Pflanzenernahrung Bodenkunde. 149: 411-427.

Kabir Z, O'Halloran I P, Fyles J W and Hamel C 1997 Seasonal changes of arbuscular mycorrhizal fungi as affected by tillage practices and fertilization: hyphal density and mycorrhizal root colonization. Plant and Soil 192: 285-293.

Kim K Y, Jordan D and McDonald G A 1998 Effect of phosphate-solubilizing bacteria and vesicular-arbuscular mycorrhizae on tomato growth and soil microbial activity. Biology and Fertility of Soils 26: 79-87.

Kjoller R and Rosendahl S 2000 (a) Detection of arbuscular mycorrhizal fungi (Glomales) in roots by nested PCR and SSCP (single stranded conformation polymorphism). Plant and Soil 226: 189-196.

Kjoller R and Rosendahl S 2000 (b) Effects of fungicides on arbuscular mycorrhizal fungi: differential responses in alkaline phosphatase activity of external and internal hyphae. Biology and Fertility of Soils 31: 361-365.

Klironomos J N, Bednarczuk E M and Neville J 1999 Reproductive significance of feeding on saprobic and mycorrhizal fungi by the collembolan, *Folsomia candida*. Functional Ecology 13: 756-761.

Klironomos J N and Moutoglis P 1999 Colonization of nonmycorrhizal plants by mycorrhizal neighbours as influenced by the collembolan, *Folsomia candida*. Biology and Fertility of Soils 29: 277-281.

Koide R T 1991 Nutrient supply, nutrient demand and plant response to mycorrhiza infection. Tansley review no. 29. New Phytologist 117: 364-386.

Koide R T 2000 Functional complementarity in the arbuscular mycorrhizal symbiosis. New Phytologist 147: 233-235.

Kurle J E and Pfleger F L 1996 Management influences on arbuscular mycorrhizal fungal species composition in a corn-soybean rotation. Agronomy Journal 88: 155-161.

Larsen J, Thingstrup I, Jakobsen I and Rosendahl S 1996 Benomyl inhibits phosphorus transport but not fungal alkaline phosphatase activity in a *Glomus* – cucumber symbiosis. New Phytologist 132: 127-133.

Li X-L, George E and Marschner H 1991 Phosphorus depletion and pH decrease at the root-soil and hyphae-soil interfaces of VA mycorrhizal white clover fertilized with ammonium. New Phytologist 119: 397-404.

Linderman R G 2000 Effects of mycorrhizas on plant tolerance to diseases. *In:* Arbuscular Mycorrhizas: Physiology and Function. Y Kapulnik and D D Douds Jr (eds.) pp. 345-365. Kluwer Academic Publishers. Dordrecht, The Netherlands.

Liu R-J 1995 Effect of vesicular-arbuscular mycorrhizal fungi on Verticillium wilt of cotton. Mycorrhiza. 5: 293-297.

Malloch D W, Pirozynski K A and Raven P H 1980 Ecological and evolutionary significance of mycorrhizal symbioses in vascular plants (a review). Proc. Nat. Acad. Sci. USA. 77: 2113-2118.

Manjunath A, Mohan R and Bagyaraj D J 1981 Interaction between *Beijerinckia mobilis, Aspergillus niger* and *Glomus fasciculatus* and their effects on growth of onion. New Phytologist 87: 723-727.

Mark G L and Cassells A C 1999 The effect of dazomet and fosetyl-aluminum on indigenous and introduced arbuscular mycorrhizal fungi in commercial strawberry production. Plant and Soil 209: 253-261.

Matsubara Y-i, Tamura H and Harada T 1995 Growth enhancement and Verticillium wilt control by vesicular-arbuscular mycorrhizal fungus inoculation in eggplant. J. Jap. Soc. Hort. Sci. 64: 555-561.

McAllister C B, Garcia-Romera I, Martin J, Godeas A and Ocampo J A 1995 Interaction between *Aspergillus niger* van Teigh. and *Glomus mosseae* (Nicol. and Gerd.) Gerd. and Trappe. New Phytologist 129: 309-316.

McGonigle T P 1988 A numerical analysis of published field trials with vesicular-arbuscular mycorrhizal fungi. Functional Ecology 2: 473-478.

McGonigle T P, Evans D G and Miller M H 1990 Effect of degree of soil disturbance on mycorrhizal colonization and phosphorus absorption by maize in growth chamber and field experiments. New Phytologist 116: 629-636.

McGonigle T P and Miller M H 1993 Mycorrhizal development and phosphorus absorption in maize under conventional and reduced tillage. Soil Science Society of America Journal 57: 1002-1006.

McGonigle T P and Miller M H 1996 Mycorrhizae, phosphorus absorption, and yield of maize in response to tillage. Soil Science Society of America Journal 60: 1856-1861.

McGonigle T P and Miller M H 2000 The inconsistent effect of soil disturbance on colonization of roots by arbuscular mycorrhizal fungi: a test of the inoculum density hypothesis. Applied Soil Ecology 14: 147-155.

McGraw A-C and Hendrix J W 1984 Host and soil fumigation effects on spore population densities of species of Endogenaceous mycorrhizal fungi. Mycologia. 76: 122-131.

Menge J A 1982 Effect of soil fumigants and fungicides on vesicular-arbuscular mycorrhizal fungi. Phytopathology 72: 1125-1132.

Menge J A, Steirle D, Bagyarai D J, Johnson E L V and Leonard R T 1978 Phosphorus concentrations in plants responsible for inhibition of mycorrhizal infection. New Phytologist 80: 575-578.

Merryweather J and Fitter A 1998 The arbuscular mycorrhizal fungi of *Hyacinthoides non-scripta*. I. Diversity of fungal taxa. New Phytologist 138: 117-129.

Meyer J R and Linderman R G 1986 Selective influence on populations of rhizosphere or rhizoplane bacteria and actinomycetes by mycorrhizas formed by *Glomus fasciculatum*. Soil Biology and Biochemistry 18: 191-196.

Miller M H 2000 Arbuscular mycorrhizae and the phosphorus nutrition of maize: a review of Guelph studies. Canadian Journal of Plant Science 80: 47-52.

Miller M H, McGonigle T P and Addy H D 1995 Functional ecology of vesicular-arbuscular mycorrhizas as influenced by phosphate fertilization and tillage in an agricultural ecosystem. Critical Reviews in Biotechnology 15: 241-255.

Miller R L and Jackson L E 1998 Survey of vesicular-arbuscular mycorrhizae in lettuce production in relation to management and soil factors. Journal of Agricultural Science 130: 173-182.

Miller R M and Jastrow J D 1990 Hierarchy of root and mycorrhizal fungal interactions with soil aggregation. Soil Biology and Biochemistry 22: 579-584.

Miller R M and Jastrow J D 1992 The role of mycorrhizal fungi in soil conservation. *In:* Mycorrhizae in Sustainable Agriculture. G J Bethlenfalvay and R G Linderman (eds.) pp. 29-44. Agronomy Society of America Special Publication No. 54. Madison, WI.

Miller R M and Jastrow J D 2000 Mycorrhizal fungi influence soil structure. *In* Arbuscular Mycorrhizas: Physiology and Function. Y Kapulnik and D D Douds Jr (eds.) pp. 3-18. Kluwer Academic Publishers. Dordrecht, The Netherlands.

Miller R M and Kling M 2000 The importance of integration and scale in the arbuscular mycorrhizal symbiosis. Plant and Soil 226: 295-309.

Mosse B 1962 The establishment of VA mycorrhiza under aseptic conditions. Journal of General Microbiology 27: 509-520.

Mosse B 1973 Advances in the study of vesicular-arbuscular mycorrhiza. Annual Review of Phytopathology 11: 171-196.

Mozafar A, Jansa J, Ruh R, Anken T, I Sanders and Frossard E 2001 Functional diversity of AMF co-existing in agricultural soils subjected to different tillage. Proceeding of the Third International Conference on Mycorrhizas. July 8-13, 2001. pp. P1 32. Adelaide, South Australia.

Ness R L L and Vlek P L G 2000 Mechanism of calcium and phosphate release from hydroxy-apatite by mycorrhizal fungi. Soil Science Society of America Journal 64: 949-955.

Newsham K K, Fitter A H and Watkinson A R 1995 Arbuscular mycorrhiza protect an annual grass from root pathogenic fungi in the field. Journal of Ecology 83: 991-1000.

Norman J R and Hooker J E 2000 Sporulation of *Phytophthora fragariae* shows greater stimulation by exudates of non-mycorrhizal than by mycorrhizal strawberry roots. Mycological Research 104: 1069-1073.

Nurlaeney N, Marschner H and George E 1996 Effects of liming and mycorrhizal colonization on soil phosphate depletion and phosphate uptake by maize (*Zea mays* L.) and soybean (*Glycine max* L.) grown in two tropical acid soils. Plant and Soil 181: 275-285.

Oger P, Mansouri H and Dessaux Y 2000 Effect of crop rotation and soil cover on alteration of the soil microflora generated by the culture of transgenic plants producing opines. Molecular Ecology 9: 881-890.

Pfleger F L and Linderman R G (eds.) 1994 Mycorrhizae and Plant Health. p. 344. American Phytopathological Society Press. St. Paul, MN.

Piccini D and Azcon R 1987 Effect of phosphate-solubilizing bacteria and vesicular-arbuscular mycorrhizal fungi on the utilization of Bayovar rock phosphate by alfalfa plants using a sand-vermiculite medium. Plant and Soil 101: 45-50.

Powell C L L and Daniel J 1978 Mycorrhizal fungi stimulate uptake of soluble and insoluble phosphate fertilizer from a phosphate-deficient soil. New Phytologist 80: 351-358.

Read D J, Francis R and Findlay R D 1985 Mycorrhizal mycelia and nutrient cycling in plant communities. *In:* Ecological Interactions in Soil. A H Fitter, D Atkinson, D J Read and M Busher (eds.) pp. 193-217. Blackwell Scientific Publications. London.

Redecker D, Kodner R and Graham L E 2000 Glomalean fungi from the Ordovician. Science 289: 1920-1921.

Ross J P 1980 Effect of nontreated field soil on sporulation of vesicular-arbuscular mycorrhizal fungi associated with soybean. Phytopathology 70: 1200-1205.

Ross J P and Ruttencutter R 1977 Population dynamics of two vesicular-arbuscular endomycorrhizal fungi and the role of hyperparasitic fungi. Phytopathology 67: 490-496.

Ruiz-Lozano J M and Azcon R 1995 Hyphal contribution to water uptake in mycorhizal plants as affected by fungal species and water stress. Physiol. Plant. 95: 472-478.

Ryan M H and Angus J F 2001 Role of VAM fungi in growth and nutrient uptake of wheat and field peas in the southern wheatbelt of Australia. Proceedings of the Third International Conference on Mycorrhizas. July 8-13, 2001. pp. C6 08. Adelaide, South Australia.

Saif S R 1986 Vesicular-arbuscular mycorrhizae in tropical forage species as influenced by season, soil texture, fertilizers, host species and ecotypes. Angew. Bot. 60: 125-139.

St-Arnaud M, Hamel C, Caron M and Fortin J A 1994 Inhibition of *Pythium ultimum* in roots and growth substrate of mycorrhizal *Tagetes patula* colonized with *Glomus intraradices*. Canadian Journal of Plant Pathology 16: 187-194.

Sanders F E, Mosse B and Tinker P B 1975 Endomycorrhizas. Academic Press. London.

Sanders I R 1999 No sex please, we're fungi. Nature 399: 737-739.

Sattelmacher B, Reinhrad S and Pomilkalko A 1991 Differences in mycorrhizal colonization of rye (*Secale cereale L.*) grown in conventional or organic biological-dynamic farming systems. Journal of Agronomic Crop Science 167: 350-355.

Schenck N C, Sigueira J O and Oliveira E 1989 Changes in the incidence of VA mycorrhizal fungi with changes in ecosystems. *In:* Interrelationships between Microorganisms and Plants in Soil. V Vancura and F Kunc (eds.) pp. 125-129. Elsevier. New York.

Schreiner R P and Bethlenfalvay G J 1997 Mycorrhizae, biocides, and biocontrol: III. Effects of three different fungicides on developmental stages of three AM fungi. Biology and Fertility of Soils 24: 18-26.

Schreiner R P, Mihara K L, McDaniel H and Bethlenfalvay G J 1997 Mycorrhizal fungi influence plant and soil functions and interactions. Plant and Soil. 188: 199-209.

Sieverding E 1990 Ecology of VAM fungi in tropical agroecosystems. Agricriculture, Ecosystems and Environment 29: 369-390.

Simon L, Bousquet J, Levesque R C and Lalonde M 1993 Origin and diversification of endomycorrhizal fungi and coincidence with vascular land plants. Nature 363: 67-69.

Sinsabaugh R L 1994 Enzymatic analysis of microbial pattern and process. Biology and Fertility of Soils 17: 69-74.

Slezack S, Dumas-Gaudot E, Paynot M and Gianinazzi S 2000 Is a fully established arbuscular mycorrhizal symbiosis required for bioprotection of *Pisum sativum* roots against *Aphanomyces eutriches*? Molec. Plant Microbe Interact. 13: 238-241.

Smith F A, Jakobsen I and Smith S E 2000 Spatial differences in acquisition of soil phosphate between two arbuscular mycorrhizal fungi in symbiosis with *Medicago truncatula*. New Phytologist 147: 357-366.

Smith T F 1978 A note on the effect of soil tillage on the frequency and vertical distribution of spores of vesicular-arbuscular endophytes. Australian Journal of Soil Research 16: 359-361.

Spokes J R, Hayman D S and Kandasamy D 1989 The effects of fungicide-coated seeds on the establishment of VA mycorrhizal infection. Annals of Applied Biology 115: 237-241.

Stahl P D and Smith W K 1984 Effects of different geographic isolates of *Glomus* on the water relations of *Agropyron smithii*. Mycologia 76: 261-267.

Stahl P D and Christensen M 1990 Population variation in the mycorrhizal fungus *Glomus mosseae*: uniform garden experiments. Mycological Research 94: 1070-1076.

Stewart C N Jr, Richards H A and Halfhill M D 2000 Transgenic plants and biosafety: science, misconceptions and public perceptions. BioTechniques 29: 832-843.

Stubblefield S P, Taylor T N and Trappe J M 1987 Fossil mycorrhizae: a case for symbiosis. Science 237: 59-60.

Sukarno N, Smith S E and Scott E S 1993 The effect of fungicides on vesicular-arbuscular mycorrhizal symbiosis. I. The effects of vesicular-arbuscular mycorrhizal fungi and plant growth. New Phytologist 25: 139-147.

Sylvia D M and Neal L H 1990 Nitrogen affects the phosphorus response of VA mycorrhiza. New Phytologist 115: 303-310.

Sylvia D M, Wilson D O, Graham J H, Maddox J J, Millner P, Morton J B, Skipper H D, Wright S F and Jarstfer A G 1993 (a) Evaluation of vesicular-arbuscular mycorrhizal fungi in diverse plants and soils. Soil Biology and Biochemistry 25: 705-713.

Sylvia D M, Hammond L C, Bennett J M, Haas J H and Linda S B 1993 (b) Field response of maize to a VAM fungus and water management. Agronomy Journal 85: 193-198.

Tarafdar J C and Marchner H 1995 Dual inoculation with *Aspergillus funigatus* and *Glomus mosseae* enhances biomass production and nutrient uptake in wheat (*Triticum aestivum* L.) supplied with organic phosphorus as Na-phytate. Plant and Soil 173: 97-102.

Thimm T and Larink O 1995 Grazing preferences of some collembola for endomycorrhizal fungi. Biology and Fertility of Soils 19: 266-268.

Thomas R S, Dakessian S, Ames R N, Brown M S and Bethlenfalvay G J 1986 Aggregation of a silty clay loam soil by mycorrhizal onion roots. Soil Science Society of America Journal 50: 1494-1499.

Thomas R S, Franson R L and Bethlenfalvay G J 1993 Separation of vesicular-arbuscular mycorrhizal fungus and root effects on soil aggregation. Soil Science Society of America Journal 57: 77-81.

Thompson J P 1987 Decline of vesicular-arbuscular mycorrhizae in long fallow disorder of field crops and its expression in phosphorus deficiency of sunflower. Australian Journal of Agricultural Research 38: 847-867.

Thompson J P 1991 Improving the mycorrhizal condition of the soil through cultural practices and effects on growth and phosphorus uptake by plants. *In:* Phosphorus Nutrition of Grain Legumes in the Semi-Arid Tropics. C Johansen, K K Lee and K L Sahrawat (eds.) pp. 117-138. Internat. Crops Res. Inst. for the Semi-Arid Tropics.

Tilman D 1988 Plant strategies and the dynamics and structure of plant communities. Princeton University Press. Princeton New Jersey, USA.

Tisdall J M 1991 Fungal hyphae and structural stability of soil. Australian Journal of Soil Research 29: 729-744.

Tisdall J M and Oades J M 1979 Stabilisation of soil aggregates by the root systems of ryegrass. Australian Journal of Soil Research 17: 429-441.

Tisdall J M and Oades J M 1982 Organic matter and water-stable aggregates in soils. Journal of Soil Science 33: 141-163.

Torres-Barragan A, Zavaleta-Mejia E, Gonzalez-Chavez C and Ferrera-Cerrato R 1996 The use of arbuscular mycorrhizae to control onion white rot (*Sclerotium cepivorum* Berk.) under field conditions. Mycorrhiza 6: 253-257.

Traquair J A 1995 Fungal biocontrol of root diseases: endomycorrhizal suppression of Cylindricarpon root rot. Canadian Journal of Botany 73(suppl): S89-S95.

Trotta A, Varese G C, Gnavi E, Fusconi A, Sampo S and Berta G 1996 Interactions between the soilborne root pathogen *Phytophthora nicotianae* var *parasitica* and the arbuscular mycorrhizal fungus *Glomus mosseae* in tomato plants. Plant and Soil 185: 199-209.

Vancura V 1986 Microbial interactions in the soil. *In:* Physiological and Genetical Aspects of Mycorrhizae. V Gianinazzi-Pearson and S Gianinazzi (eds.) pp. 189-196. INRA publishing service. Versailles, France.

Vigo C, Norman J R and Hooker J E 2000 Biocontrol of the pathogen *Phytophthora parasitica* by arbuscular mycorrhizal fungi is a consequence of effects on infection loci. Plant Pathology 49: 509-514.

Wacker T L, Safir G R and Stephens C T 1990 Effect of *Glomus fasciculatum* on the growth of asparagus and the incidence of Fusarium root rot. J. Amer. Soc. Hort. Sci. 115: 550-554.

Warnock A J, Fitter A H and Usher M B 1982 The influence of a springtail *Folsoma candida* (Insecta, Collembola) on the mycorrhizal association of leek (*Allium porrum*) and the vesicular-arbuscular mycorrhizal endophyte *Glomus fasciculatus*. New Phytologist 90: 285-292.

West H M, Fitter A H and Watkinson A R 1993 Response of *Vulpia ciliata* ssp. *ambigua* to removal of mycorrhizal infection and to phosphate application under natural conditions. Journal of Ecology 81: 351-358.

Wetterauer D G and Killorn R J 1996 Fallow- and flooded-soil syndromes: effects on crop production. Journal of Production Agriculture 9: 39-41.

Wilson G W T, Hetrick B A D and Gerschefske D K 1988 Suppression of mycorrhizal growth response of big bluestem by non-sterile soil. Mycologia 80: 338-343.

Wright S F, Franke-Snyder M, Morton J B and Upadhyaya A 1996 Time-course study and partial characterization of a protein on hyphae of arbuscular mycorrhizal fungi during active colonization of roots. Plant and Soil 181: 193-203.

Wright S F and Upadhyaya A 1996 Extraction of an abundant and unusual protein from soil and comparison with hyphal protein of arbuscular mycorrhizal fungi. Soil Science 161: 575-586.

Wright S F and Upadhyaya A 1998 A survey of soils for aggregate stability and glomalin, a glycoprotein produced by hyphae of arbuscular mycorrhizal fungi. Plant and Soil. 198: 97-107.

Wright S F, Starr J L and Paltineau I C 1999 Changes in aggregate stability and concentration of glomalin during tillage management transition. Soil Science Society of America Journal 63: 1825-1829.

Wright S F and Anderson R L 2000 Aggregate stability and glomalin in alternative crop rotations for the central Great Plains. Biology and Fertility of Soils 31: 249-253.

Ye X, Al-Babili S, Klöti A, Zhang J, Lucca P, Beyer P and Potrykus I 2000 Engineering the protovitamin A (B-carotene) biosynthetic pathway into (carotenoid-free) rice endosperm. Science 287: 303-305.

Chapter 8

Relevance of Plant Root Pathogens to Soil Biological Fertility

V.V.S.R. Gupta[1] and K. Sivasithamparam[2]

[1] CSIRO Land and Water, Glen Osmond, SA, 5064, Australia
[2] School of Earth and Geographical Sciences, Faculty of Natural and Agricultural Sciences, The University of Western Australia, Crawley, 6009, Western Australia

1. INTRODUCTION

In this chapter we use the term soil biological fertility to describe the ability of soil biota to perform various (1) plant essential functions to support the growing plant with its nutritional and other biological requirements, and (2) ecosystem functions that maintain the quality of soil resource. A number of soil functions essential for plant growth and crop productivity are regulated by different groups of biota. These include (i) mineralisation and uptake of major nutrients (e.g. N, P and S) and trace elements (e.g. Zn), (ii) beneficial, pathogenic and associative interactions affecting root and shoot growth, (iii) degradation of chemicals harmful for plant growth (e.g. herbicides from a previous cropping season), and (iv) formation of soil structural components that provide optimal aeration and water-filled pore space for plant growth.

163

L.K. Abbott & D.V. Murphy (eds) Soil Biological Fertility - A Key to Sustainable Land Use in Agriculture. 163-185
© 2007 Springer.

A unique balance between the three components of a soil system, i.e. physical, chemical and biological, is necessary for long-term sustainability of crop production, soil health and other essential ecosystem functions. Soil biota regulate processes that impact on the physical and chemical properties of soil and conversely the physical and chemical attributes of soil greatly influence the populations and activities of soil biota. The optimum functioning of the biological components of soil requires both a suitable habitat (pH, habitable pore space, oxygen concentration etc.) and optimum environmental conditions (temperature, moisture level etc.). For example, the activities of different groups of soil biota have important roles in various components of soil structure i.e. the burrowing activities of macrofauna influencing soil pore structure (Lee and Foster 1991) and binding and entanglement of soil particles by microflora (including pathogenic fungi) in the aggregate formation and stabilisation (Gupta and Germida 1988, Tisdall 1991, Tisdall *et al.* 1997). Conversely stable aggregates are an important component of soil structure for maintaining aeration and porosity for favourable microbial growth including that of plant pathogenic fungi. These activities may affect the physical and chemical properties, which are known to determine suppressiveness of soils to plant root diseases (Hoper and Alabouvette 1996).

Management practices involving surface retention of crop residues are recommended for improving soil organic matter, soil structure and biota populations. However, they can result in providing food source for the survival of some pathogenic microorganisms especially during off-season. In addition, the surface retention of residues has increased the potential for the movement of residues around the farm and across farms and might have increased the carryover of soilborne root pathogens (Neate 1994, Allen and Lonergan 1998). Allen (2000) summarised the trends in diseases in Australian cotton based on 17 years of survey data on disease incidence and severity. Results show that there has been a steady decline in the mean seedling mortality during the 10-year period ending in 1998, i.e. from 50% mean seeding mortality in 1987/88 to <30% seedling mortality in 1997/98. The incidence of diseases such as bacterial blight (*Xanthomonas campestris*) and verticillium wilt (*Verticillium dahliae*) has declined whereas diseases such as black root rot (*Thielaviopsis basicola*) and fusarium blight (*Fusarium oxysporum* f.sp. *vasinfectum*) have become a severe threat to the sustainability of Australian cotton industry.

Presence of the inoculum of a pathogen does not necessarily result in the outbreak of the disease and the severity of the disease is ultimately determined by the environmental conditions. Irrigated wheat often succumbs to pathogens different to rainfed crops. In the Pacific northwest (Cook and Baker 1983) fusarium crown rot dominated rainfed crop while take-all (caused by *Gaeumannomyces graminis* var. *tritici*) was the major problem in irrigated wheat. *Fusarium* dominates in soils that are relatively dry and with relatively low microbial (mainly bacterial) activity, while the take-all fungus although not as saprophytically competent as *Fusarium* spp., is active in moist soils. The extent of threat from major pathogens varies with crops and regions of Australia (Murray and Brennan 2001, unpublished). For example, Fusarium crown rot of wheat, especially that caused by *F. pseudograminearum*, is predominantly a problem in the north and central cropping regions of Australia, while the crops in the south central wheat growing region are more severely affected by pathogens such as *Rhizoctonia solani* AG8 and take-all fungus (*Gaeumannomyces graminis* var. *tritici*).

2. NATURE OF RELATIONSHIP BETWEEN PLANT ROOT PATHOGENS AND SOIL BIOLOGICAL FERTILITY

2.1 Interactions between Root Health and Nutrient and Water Uptake by Crops

Soil borne plant pathogens affect biological fertility directly and indirectly. Directly, they affect the efficiency of root's capacity to acquire water and nutrients. Indirectly, the reduction in plant biomass resulting from disabling the host leads to reduced input of the quality and quantity of organic matter that eventually enters the soil. Neate (unpublished data, personal communication) working with rhizoctonia bare patch soils in Spalding, South Australia observed that the unused water by the wheat plants in the rhizoctonia bare patch areas exceeded 50mm, also there was 56 kg N /ha unused nitrate nitrogen in the surface one meter soil profile. The 50 mm unused water accounts for more than 25% of growing season rainfall. Deep leaching of this unused water and mineral nitrogen below root zone could lead to a number of environmental problems and degradation of landscape at a large scale. Inadequate usage of available water by the disease-affected crops is one of the critical reasons for the increased drainage under dryland agricultural crops. Thus, plant pathogens and plant disease management play an important role in the fertility of the cropped land and also the health of the whole agroecosystem.

The value of soil fertility to the plant is only of value if plants can access this fertility. Root health and vigour is of critical importance to the nutrition of the plant. With wheat for example, under the Mediterranean conditions of southern Australia, the five seminal roots it produces are critical especially in accessing stored water. They are the only root system available for the first 21 days of the seedling's life and hence are targeted by most soil-borne root pathogens. Early infections debilitate the young plant which becomes nutrient impoverished because of dysfunctional roots. This predisposes the plant to additional attacks by other necrotrophic plant pathogens, which favour weak, nutrient-deficient plants.

In any given environment, for each farming system, a dynamic equilibrium exists between disease severity, soil biological fertility and plant production. A change in one or more components, in particular the one most limiting, may be required for any production system that is performing below optimum in order to improve productivity and resource use efficiency. Management of disease-affected crops by merely targeting the pathogen attack is unlikely to be adequate. For instance, in the wheat fields on highly calcareous soils of the Eyre Peninsula in South Australia where biological fertility of soils, especially relating to root growth and nutrient availability are limiting, just tackling the pathogen alone is not likely to significantly improve crop production until other major nutrient and biological constraints related to poor soil fertility are overcome.

2.2   Interactions between Beneficial Bacteria and Root Pathogens

Rhizobacteria capable of increasing shoot and root growth through a number of different mechanisms have been described both for dryland and irrigated agricultural crops and horticultural plants. These bacteria, known as plant growth promoting rhizobacteria (PGPR), are considered to improve plant growth either through biocontrol of plant pathogens or by increasing root and shoot growth both in the presence or absence of a disease. Some PGPR strains have also been shown to induce systemic resistance against multiple pathogens including bacteria, fungi, viruses and nematodes (Kloepper *et al.* 1999, 2000). The development of an integrated disease management approach which incorporates both the PGPR and induced systemic resistance (ISR) has greater chance of success for consistent disease suppression of both major and minor pathogens. *Rhizobium* spp. have also shown excellent potential as plant growth promoting rhizobacteria (PGPR) with non-legumes (Antoun *et al.* 1998). On colonisation of non-legumes they may produce phytohormones, siderophores and HCN. Some strains of rhizobia are also antagonistic towards some plant pathogenic fungi. *Rhizobium* may also directly compete with fungal pathogens resulting in the reduced severity of root-rots (Tu 1978).

Most of the research on beneficial effects of biocontrol microorganisms and other PGPR has been done with organisms that have been selected for effective performance in rhizosphere environments. Recently, there has been scientific and commercial interest in the use of endophytic microorganisms.

Endophytes, because of their intimate association with their host plants, their location and action, avoid environmental adversities that exist in the rhizosphere and bulk soil (Sivasithamparam 1998). However, rhizosphere microorganisms capable of endophytic activity could confer a distinct benefit to the host plant against a pathogen before it reaches the plant itself, thus providing a dual control mechanism (Kobayashi and Palumbo 2000, Sturtz *et al*. 2000). A variety of stimuli, both biotic and abiotic, appear to induce resistance against a wide spectrum of plant pathogens (Van Loon *et al*. 1998). In many cases the induced resistance appears to be effective against a broad spectrum of targeted pathogens.

2.3   Interactions between Mycorrhizae and Root Pathogens

In addition to the nutritional benefits (e.g. P and Zn) from mycorrhizal colonisation, mycorrhizal plants are also known to better withstand adverse environmental conditions, such as drought and salinity, due to the increased accessibility of soil water by the mycorrhizal roots and hyphal networks (Davies *et al*. 1992, Subramanian and Charest 1999). Another important benefit from mycorrhizal symbiosis is the protection from plant pathogenic fungi (Fernando and Linderman 1997) and nematodes. Although a cascade of biochemical changes occur in roots colonised by mycorrhizal fungi (Smith and Read 1997) it is likely that the reduction in root diseases in mycorrhizal plants is predominantly due to improved nutrition and vigour of the plant host. The improved phosphorus status of the wheat plant leading to a decrease in the net leakage of root exudates and thereby reduced pathogen growth in the rhizosphere has been suggested as the basis of reduced take-all severity in wheat plants inoculated with mycorrhizal fungi (Graham and Menge 1982).

2.4   Interactions between Soil Fauna and Root Pathogens

Soil fauna influence plant disease incidence through i) their effects on the survival, growth and transportation/dispersal of pathogen inoculum including fungal propagules, ii) host plant-pathogen interaction and, iii) the nutrition of the host plant (Lussenhop and Wicklow 1984, Old 1986, Curl 1988, Gupta 1994). The presence of high levels of mycophagous amoebae has been associated with disease suppression or reduced disease caused by take-all fungus, verticillum wilt fungus and *Rhizoctonia* causing bare-patch, in laboratory and glasshouse trials (Old 1986, Gupta *et al*. 1996). Mesofaunal grazing selectively on pathogenic fungal hyphae (e.g. Collembola grazing on *R. solani*; Lartey *et al*. 1989) and altering of rhizosphere fungal community structure has also been reported (Curl *et al*.

1983). Such reports also show the effects of macrofauna on pathogenic fungi. However, most of these reports are based on controlled environment studies and limited evidence is available showing direct relationships under field conditions. In general, management systems that support high and diverse soil faunal communities do not favour the proliferation of certain pathogenic fungi. Gupta *et al.* (1995) observed, in field based studies, that the reduction in the survival of *Rhizoctonia solani* and take-all fungal inoculum was associated with high populations of mycophagous amoebae and fungal feeding nematodes (mycophagous effects), suggesting that they could have a significant role in the broad spectrum disease suppression observed in a long-term farming systems trial at Avon, South Australia (Roget 1995).

Predation of microflora by soil fauna, in particular microfauna, releases plant available nutrients tied up in microorganisms. Better nutrition of plants could result in reduced overall negative effects from plant disease incidence. Gupta *et al.* (1999) observed that wheat plants grew better in soil cores even after inoculation with *R. solani* fungus in the presence of mesofauna. High populations of mesofauna resulted in higher levels of nutrients in plants (eg concentration and total uptake of nitrogen) suggesting that better nutrition in the presence of mesofauna could be a reason for reduced *Rhizoctonia* disease severity.

3. PLANT PATHOGENS AS COMPONENTS OF THE SOIL FOOD-WEB

Many biological processes in terrestrial ecosystems are mediated or regulated by multiple species or trophic groups of organisms. Therefore, successful functioning of most soil biological processes requires a balance of interactions in a complex soil biota community (detritus food web). In a detritus food web, organisms across trophic levels are linked on the basis of the flow of energy and food preference. While simple microbial-faunal interactions have been used to explain the effectiveness of an introduced biocontrol microorganism, complex food-web structures are needed to delineate mechanisms or predict changes in biological functions such as crop residue decomposition, nutrient (N, S and P) cycling and disease suppression. In a detrital food web, microflora including beneficial and pathogenic bacteria and fungi form the primary decomposer groups that transfer the carbon and nutrients from crop residues into the soil biota component. A conceptual framework indicating the various groups of soil biota and their linkages along with the different biological functions they might influence, based on published information, is given in Figure 1.

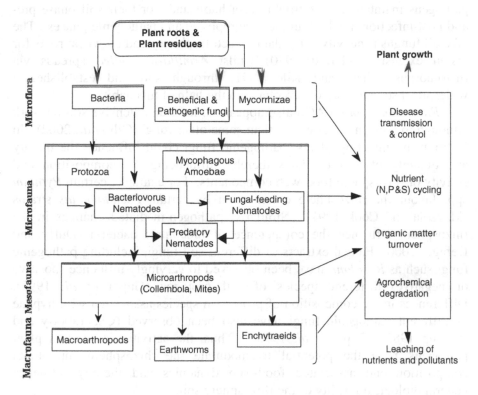

Figure 1 The different groups of soil biota are linked in a detritus food-web model in order to express their role in key soil biological processes. This model is based on published information (based on information from Hendrix *et al.* 1986, Beare *et al.* 1995, Roper and Gupta 1995 and Gupta and Neate 1999).

Carbon compounds from root exudates provide the energy source for the rhizosphere food-web complex whereas decomposing crop residues form one of the critical microsites for the food-web community in the bulk soil. Predation by protozoa and nematodes (microfauna) on microflora can lead to the release of plant available forms of nitrogen, phosphorus and sulfur and contributes to a significant portion of nutrient uptake by plants (Gupta and Yeates 1997). Predation by microfauna can also modify the community structure of bacteria and fungi in the rhizosphere. As bacteria are a major food source for microfauna, bacterial pathogens in the rhizosphere of infected plants could also become a major component of the primary food source for the higher trophic levels. Thus, these pathogens may play a significant role in the rhizosphere processes. Similarly, pathogenic fungi may form a major component of the rhizosphere food-web due to their external hyphae both pre- or post infection of a host plant. Many root pathogens maintain an ectotrophic root habit and / or their soil phase pre- and post-infection, i.e. both in their saprophytic and pathogenic phases. The take-all fungus is heavily dependant on ectotrophic spread on wheat roots for lesion extension (Garrett 1970) whilst *Armillaria mellea* spreads via rhizomorphs (Shaw and Kile 1991) through soil and establishes a widespread soil network of mycelia with the help of this habit.

Rhizoctonia solani AG8 also appears to have an extensive soil network although to a much lesser extent than some other fungi (Gill *et al.* 2002). In addition, by infecting plants, pathogenic fungi could influence the quality and quantity of root exudates thereby influencing the composition and activity of rhizosphere food-web components. For example, certain *Pythium* spp. favour the rhizosphere colonisation by certain *Pseudomonas* strains (Mazzola and Cook 1991). Similarly, pathogen-mediated changes in the rhizosphere influence the competence of biocontrol bacteria (Duffy and Defago, 2000). Fungal extracts of different soil fungi, including pathogenic fungi such as *R. solani,* have been observed to varyingly influence (positive or negative) different species of soil protozoa (Gupta *et al.* 1995). Differences in the composition of protozoan species associated with hyphae of different pathogenic fungi have also been observed (Chakraborty and Warcup 1985, Gupta *et al.* 1995). Thus, the activities of certain plant pathogens have the potential to modulate the rhizosphere microbiota composition and associated food-web dynamics and thereby influence general biological fertility of the rhizosphere soil.

4.   PLANT PATHOGENS AND SOIL BIOTA IN THE PRESENCE OR ABSENCE OF A PLANT

Some pathogenic microorganisms participate in general soil biological functions, e.g. decomposition, nutrient cycling and soil aggregation, in their saprophytic phase especially in the absence of their specific host plant. This could happen during the off-season or under rotations with non-hosts. Such beneficial functions may be significant for pathogenic fungi and bacteria but are not applicable for organisms such as plant parasitic nematodes. Does this mean the beneficial functions of pathogenic microorganisms have hitherto been under-rated in relation to their role in soil biological fertility?

Soil-borne necrotrophic fungal pathogens may vary in their saprophytic competency. The take-all fungus, a pathogen considered to have a relatively low level of competitive saprophytic ability (Garrett 1970) depends heavily on its survival in the residues of cereal roots and crowns colonised during its pathogenic phase. *Rhizoctonia solani,* on the other hand, may also depend on the hyphal network in soil, which is essential for disease establishment

(MacDonald and Rovira 1985). Gill *et al.* (2002) demonstrated that the area of the bare-patch of wheat caused by *R. solani* AG8 is determined by the extent to which the mycelial network is established prior to sowing. This may explain the sensitivity of this pathogen to cultivation prior to seeding. Crop residues are important microsites for the survival during off-season for the cotton pathogen *Fusarium oxysporum* f.sp. *vasinfectum* (Allen and Lonergan 1998). In contrast to the take-all fungus, *Fusarium* and *Rhizoctonia* fungi survive well outside infected plant tissues, as they establish hyphal networks in the bulk soil as well as colonise crop residues i.e. high degree of saprophytic competency. This suggests that these fungi could form a significant proportion of fungal biomass in the soil following a susceptible host crop.

Saprophytically competent pathogens (e.g. *Fusarium* and *Rhizoctonia solani*) can thus compete with non-pathogenic soil microbes for colonisation and mineralisation of organic matter. These fungi have been found to have relatively high cellulose adequacy indices (Garrett 1970). Hyphal networks of pathogenic fungi are thought to facilitate the formation of water stable aggregates. An increase in the number of aggregates >50 μm diam. due to the saprophytic growth of the pathogen *R. solani* is shown in Figure 2 (Tisdall *et al.* 1997). Decomposing crop residues and associated fungal networks support the formation of water stable aggregates which would help reduce the loss of surface soil due to wind erosion. Potential benefits from such microbial functions may reduce soil degradation, maintain or improve the quality of the soil resource and reduce environmental problems associated with wind erosion in southern Australian environments.

Another important point in evaluating the relevance of plant pathogens in the absence of their specific or major host plant is their ability to grow on rotation crops either as a minor pathogen, asymptomatic cortical coloniser or as general rhizosphere microorganisms. Pathogens of grain legumes can be minor pathogens on pasture legumes such as clovers and medics, e.g. *Fusarium, Pythium* spp., *Phoma medicaginis* (field peas and medics) and *Rhizoctonia solani*. This poses a problem in planning rotations which are commonly expected to reduce disease hazards (Sivasithamparam 1993). *Rhizoctonia solani* AG 8 and the take-all fungus can attack pasture grasses after a cereal crop. It is common to see a dominance of ryegrass in wheat crops affected by take-all early in the season. This could be due to low virulence of the cereal pathogen on the grass (Nilsson 1969) or to the ability of certain wheat attacking strains of the pathogen to promote the growth of rye-grass (Dewan and Sivasithamparam 1990). This may at least partly explain the dominance of rye-grass in take-all patches. Although considered to be a weed-species within crops, this grass is an important component of pastures that form a common rotation in southern Australian farms, which rely both on crops as well as sheep to be economically viable. If the grass pathogens reduce the growth of pasture grasses they may help pasture legumes to out-compete grasses for resources (e.g. water), thus enhancing the benefits from nitrogen fixation by pasture legumes. However, these benefits may not be useful if the diseased pasture grasses maintain or increase the pathogen inoculum in the non-wheat season, resulting in higher disease incidence and severity in the following crop. Therefore, when evaluating the relevance of plant pathogens to soil biological fertility, it is necessary not only to view their role in the presence of the specific host plant but also consider their importance during the periods of host plant absence.

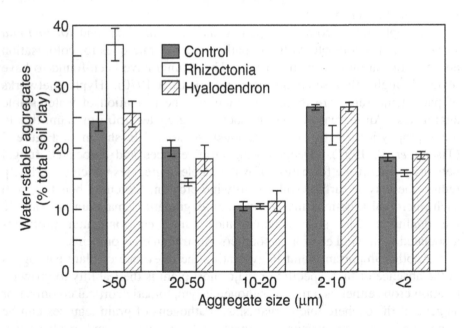

Figure 2 Effect of the saprophytic growth of fungi (for 15 days) on the size distribution of water-stable aggregates in Wiesenboden soil clay. Each error bar represents 2 x s.e.m. (from Tisdall *et al.* 1997).

5.   MANAGEMENT EFFECTS ON PLANT PATHOGENS AND CONSEQUENCES TO SOIL BIOLOGICAL FERTILITY

In recent years the use of disease break crops (e.g. broadleaf crops) along with the availability of selective grass herbicides has resulted in significant increases in wheat yields, both productivity and profitability, in

Australia, USA and Canada. In addition, during the last 10-15 years, Australian farms have also seen a general reduction in the number of cultivations per season, specialised tillage practices to reduce specific diseases such as rhizoctonia bare patch (Roget *et al*. 1996), retention of crop residues and the use of fungicides to control plant diseases. Burning crop residues and intensive cultivation of Australian agricultural soils has resulted in significant decline in soil organic matter levels including nitrogen and phosphorus concentrations (Dalal 1997, Dalal and Chan 2001) and has contributed to the loss of surface soil from severe wind erosion events. Retention of crop residues instead of burning them and a reduction in number of cultivations (generally known as conservation tillage practices) improve soil organic matter levels and reduce soil erosion. Significant improvements in total biological activity, microbial biomass as well as populations and activities of different groups of soil biota as a result of crop residue retention and reduced tillage practices have been reported in soils from different agroecological zones in Australia (Mele and Carter 1993, Pankhurst *et al*. 1995, Roper and Gupta 1995).

The widespread implementation of reduced tillage practices has been made possible by the availability of specialised herbicides, e.g. sulphonyl urea (SU) herbicides, for weed management in broadacre cereal crops, in particular in the dryland agricultural regions of southern Australia. However, both the reduced tillage and the use of SU herbicides has resulted in an increase in the incidence of rhizoctonia bare patch disease. *Rhizoctonia solani*, with its high saprophytic competency and ability to form hyphal networks, thrives well in reduced till systems, in particular with the availability of crop residues under residue retention systems. This increase in rhizoctonia bare patch has been a major limitation to the widespread implementation of conservation till systems in southern and western Australian agricultural regions. Thus, residue retention coupled with reduced cultivation provide ideal conditions for the survival and proliferation of pathogenic fungi such as *Rhizoctonia solani*. Even though stubble retention in the short-term results in an increase in the biological activity in the low fertility agricultural soils, any changes in the composition of biological community i.e. microflora and soil fauna, may require long term implementation of these practices. Roget (1995) reported the development of a broad based disease suppression phenomenon in a long-term farming system trial in South Australia. This disease suppression phenomenon was observed across different crop rotation and tillage treatments that retained crop residues and took 3-4 years to develop and a further 7 years for complete disease suppression (Figure 3). The observed disease suppression was biological in nature and active against a number of diseases including rhizoctonia root-rot, take-all, fusarium root rot (Wiseman *et al*. 1996). These observations suggest that any benefits that rhizoctonia fungi received from reduced tillage were counteracted by the changes in overall biota composition (biodiversity) when carbon supply was maintained over long periods. These soils also became suppressive to the take-all fungus over the same period (David Roget, CSIRO, personal communication).

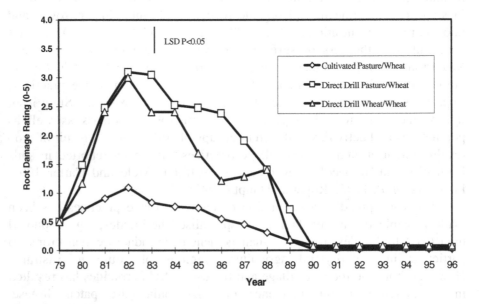

Figure 3. The development of disease suppression as indicated by the line in the rhizoctonia root rot of wheat at Avon, South Australia during 1979 to 1996 (adapted from Roget, 1995).

The nutrient-impoverished soils of southern Australia are highly conducive to diseases caused by necrotrophic plant pathogens. Rotations and fertiliser amendments to rectify these deficiencies will not only reduce the severity of diseases but also enhance the general fertility of the soils. In addition to increased yields, fertiliser applications that increase the inputs of carbon from roots and through crop residues also benefit soil biological activity. However, some negative effects from the application of specific fertilisers have been reported. For example, application of ammoniacal N suppresses take-all but leads to the acidification of bulk soil. Acidification may render the soil unsuitable for much of the bacterial activity necessary for nutrient turnover. Application of lime to enhance microbial activity in soil however renders the soil conducive to take-all (Simon and Sivasithamparam 1989).

Thus not all management practices that are recommended for improving soil biological fertility are useful to control plant diseases and vice versa. These differences in the response of specific groups of soil biota are also influenced by soil type and environmental conditions. For example, in the light textured soils in low rainfall regions of southern Australia, summer rainfall events that support higher levels of microbial activity result in the reduction of take-all fungus inoculum including those under residue retention systems (Sivasithamparam 1993, Roget 2001). In these carbon impoverished soils most of the general microbial activity including that of the take-all fungus is associated with the same microsite, i.e. fresh crop residues, and thus the take-all fungus that is resident in it is subjected to intense competition and predation by soil fauna. However, this may not apply to the heavier soils (red brown earths and heavy clay soils) in the summer rainfall regions of southeastern Australia because of the existence of adequate numbers of microsites to support both the pathogenic fungus and other microorganisms. The environment in these regions favours *Fusarium* and common root rot but not take-all.

Gupta and Neate (1999) discussed a conceptual framework which includes the factors that influence, at various stages in the pre-crop and crop growing season, the survival of pathogen inoculum from one season to the next and its effectiveness in causing disease on the susceptible plant (Figure 4). At each stage of pathogen survival and disease development, both specific organisms (micro- and macro) and general interactions between different groups are involved. For example, during the off-season (or in the absence of a suitable host plant), in order to survive in and near the substrate, the pathogen has to successfully compete (for carbon and nutrients) with other saprophytes and withstand predation by micro-, meso- and macro fauna (Gupta *et al.* 1996, Curl 1998). The distribution of microorganisms in soil is patchy; they are clumped near carbon and nutrient rich locations such as plant roots, decomposing crop residues and in micropores accessible to soluble organics. The environment is extremely heterogeneous. Pathogenic fungi have to withstand predation and competition in the carbon-poor bulk soil during their growth from the inoculum base to the host plant root. Similarly, for effective use of antagonists of pathogenic fungi (e.g. biocontrol organisms), their interaction with other microflora and predators need to be thoroughly understood (van Veen and Heijnin 1994, Bowen and Rovira 1999).

These complex biotic interactions which the pathogen needs to negotiate, necessitates the application of an integrated ecological approach (combination of functional and trophic groups and utilising the food-web model) in order to understand the mechanisms behind the reduced disease expression in disease suppressive soils. This approach may also help the development of management practices that allow the transfer of disease suppression to other soils and environments.

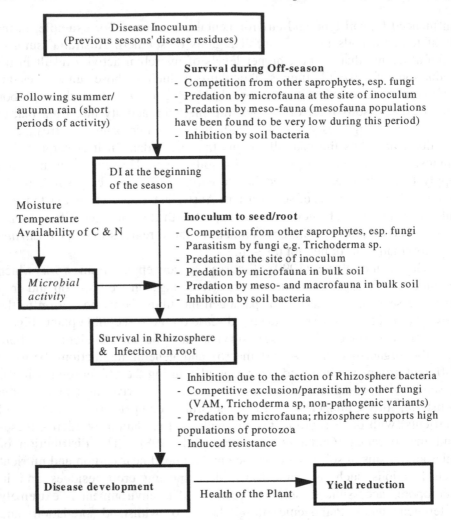

Figure 4 A conceptual frame work indicating the various factors that influence the survival and effectiveness of soil-borne root disease inoculum from one season to the next (from Gupta and Neate 1999).

6.   MANAGEMENT OF SOIL BIOTA AND BIOLOGICAL FERTILITY WHICH INFLUENCES THE ROLE OF PLANT PATHOGENS IN AGRICULTURAL SYSTEMS

In the mediterranean regions of Australia, minimum tillage and stubble retention practices have led to not just the conservation of inoculum of some plant root pathogens but also to the concentration of inoculum of necrotrophic foliar pathogens at the surface of soil. Such concentration leads to early and severe seedling infections of both shoots and roots.

One of the reasons put forward for the introduction of legumes in rotation with cereals in southern Australia has been its potential to be a 'break crop' capable of reducing the carry-over of inoculum of some cereal pathogens. While this may be true of the take-all pathogen, which attacks only graminaceous hosts, it may not apply to necrotrophic pathogens such as *Pythium* spp. and *Rhizoctonia solani* AG8. In the nitrogen deficient soils of southern Australia, legumes in rotation provide the nitrogen necessary for the soil survival of fungal pathogen attacking cereals (Garrett 1970).

To conclude, many plant pathogenic fungi in their saprophytic phase may perform a number of soil functions such as the decomposition of organic matter, mineralisation - immobilisation of nutrients, degradation of agrochemicals and soil aggregation. For example, the hyphal networks of *Rhizoctonia solani* could play a beneficial role in soil aggregation during its saprophytic stage on crop residues and in bulk soil e.g. in the absence of a host plant (Tisdall *et al.* 1997). Similarly, pathogenic fungi colonising diseased crop residues play a key role in residue decomposition and the turnover of carbon and nutrients. Such activity may play a useful role in the general soil biological fertility in the absence of a host plant i.e. off-season and in the presence of alternate crops. However, these perceived benefits may not be significant for the overall soil biological fertility of the ecosystem as plant diseases have negative effects on the key source of carbon (energy) inputs through crop residues for soil biological activity. In addition, proliferation of pathogenic fungal inoculum in the absence of a host plant is not beneficial as it may lead to higher levels of disease incidence when the host plant returns. The negative effects of plant pathogens on the grain yield, above ground and below ground biomass production, carbon inputs to the soil and reduced use of water and nutrients, finally result in soil and environmental problems within the ecosystem. These generally outweigh any beneficial role of pathogenic fungi in soil functions, in particular in Australian soils low in organic matter.

7. A MEASUREMENT OF PLANT ROOT PATHOGENS AS AN INDICATOR OF SOIL BIOLOGICAL FERTILITY

The term soil health is generally referred to 'the continued capacity of soil to function as a vital living system, within ecosystem and landscape boundaries, to sustain biological productivity, promote the quality of air and water environments, and maintain plant, animal and human health' (Doran and Safley 1997). In all definitions of soil health its ability to perform plant essential functions and productivity are considered an integral part of the concept of soil biological health. Much of the literature discussing the potential indicators of soil health generally includes beneficial soil fungi such as those which form mycorrhiza, and basidiomycetes (as pollution indicators), but pathogens however are rarely considered. Hornby and Bateman (1997) discussed in detail the advantages and limitations for using plant root pathogens as bioindicators of soil health and Pankhurst *et al.* (1997) suggested that the presence of disease might indicate the existence of a major constraint to productivity and biological fertility of soil.

An ideal indicator of soil health, including soil biological fertility, should i) be linked with ecosystem processes, ii) integrate various components, iii) respond to management and climate at an appropriate time scale, iv) be easy and cost effective to measure, and v) work well in a broad spectrum of agricultural environments. Plant pathogen related measurements e.g. the level of pathogen inoculum and the incidence of plant disease, are linked to one of the principal ecosystem processes i.e. plant growth and productivity (Hornby and Bateman 1997, Pankhurst *et al.* 1995). The level of disease incidence and severity does integrate various soil, plant and environment related factors such as the amount of pathogen inoculum, plant nutrient status and plant-beneficial biota-pathogenic fungi interactions. The amount of inoculum that reaches the plant is influenced by various soil and environmental factors e.g. pH, structure, moisture etc. Thus, the level of plant disease is an integrative indicator of different plant-soil-environmental components. In addition to being integrative, plant disease incidence and severity responds to soil management and climate in time scales that are relevant to land users and thus may be one of the useful indicators of soil biological fertility i.e. one of the members of a 'minimum data set' to evaluate soil biological fertility.

Pathogen levels in soil are dynamic and assays of pathogen at one time of the year may not reflect correctly the biological fertility or biological health of soil at other times of the season or year. Some pathogens are host specific and their surviving inoculum status during the rotation period may not be significant. For example, the inoculum status in the stubble of *Diaporthe toxica* in the absence of a lupin crop or that of *Leptosphaeria maculans* in the absence of a canola crop may not pose a threat to the crop in rotation. Even within the cropping season, the soil-borne inoculum status of damping-off pathogens (e.g. species of *Pythium*, *Rhizoctonia* etc.) pose little hazard to the crop once the plants have passed the seedling stage. Thus the selection of a single pathogenic organism as an indicator may not be possible and site or crop specific pathogen selection may be necessary. Since one of the important criteria for an ideal bioindicator is its suitability or usefulness in different environments, not all plant pathogen measurements meet this essential property. Even though pathogen measurements lack this universal applicability they may be successfully used in specific environments. Following evaluation of field environmental and disease data, covering over 25 years, Roget (2001) proposed a model for pre-season prediction of potential losses from take-all disease, utilising a DNA based assay to quantify *Gaeumannomyces graminis* var. *tritici* (Herdina and Roget 2000) coupled with the rainfall and crop rotation data.

In Figure 5 we describe a conceptual scheme indicating the various factors to consider when discussing the relevance of plant pathogens to evaluate soil biological fertility. This scheme is based on the current knowledge of the interactions among environment, management practices, plant, soil biota and pathogenic organisms since disease incidence is not just an interaction between the host plant and pathogen. The presence of plant pathogenic microorganisms alone does not result in the disease incidence even in a susceptible host plant in natural soil environments.

Soil and environmental factors that influence the pathogen reaching the host plant root and establish disease affect the usefulness of pathogen assays as bioindicators. With take-all for instance, the environment decides the extent of disease even where the resident inoculum levels are high (Cotterill and Sivasithamparam 1989). Pathogen level at any particular time of the season is not only influenced by the previous presence of a susceptible host but also by the environmental factors prior to the measurement which may have had a significant influence. For example, Roget (2001) indicated the role of summer rainfall, both amount and time of occurrence, on the level of take-all fungus at the start of the crop season in the Southern Australian soils. The influence of different crop and soil management practices on plant pathogen levels and incidence of various diseases is well documented (Rovira 1986, Neate 1994). Similarly rhizosphere interactions between beneficial and pathogenic microorganisms have a significant influence on the establishment of disease (Rovira *et al.* 1990, Bowen and Rovira 1999). Soil fertility levels (availability of N, P, K, Zn and Mn) often have greater effects on the severity of root diseases rather than on disease incidence (Huber 1981, Cotterill and Sivasithamparam 1989, Thongbai *et al.* 1993, Wilhelm *et al.* 1990).

Disease suppressiveness of soil is the ability of a soil to suppress disease severity even in the presence of a pathogen, host plant and favourable climatic conditions (Baker and Cook 1974, Simon and Sivasithamparam 1989, Roget 1995, Lemanceau *et al.* 2000). The different types of disease suppression mechanisms are related to the establishment of the pathogen, reduced parasitic activity of the pathogen and the level of disease incidence or severity. Roget *et al.* (1999) found that all soils have some potential for disease suppressiveness and thus disease suppression is not an absolute characteristic but a continuum from highly suppressive soils to poorly suppressive (i.e. conducive) soils. This means that theoretically soils could all be ranked according to their level of suppressiveness. However, such a ranking should also consider the type of suppressiveness in order to evaluate its usefulness across diseases, crops, soil types and environments.

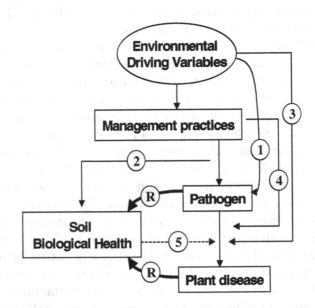

Figure 5 A conceptual scheme showing the various influencing factors in considering the relevance of plant pathogens (R) to evaluate soil biological fertility (based on reports by Neate 1994, Roper and Gupta 1995, Roget 1995, Hornby and Bateman 1997, Gupta and Neate 1999, Herdina *et al*. 2001). 1- environmental variables affecting the pathogen, effect of summer rainfall on take-all fungus inoculum (Roget 2001); 2 - management effects on general soil biological health independent of pathogen, might even be conflicting, direct drill benefits on microbial activity but negative for rhizoctonia bare patch; 3 - environmental changes affecting host response to disease; 4 - management effects that influence host response to disease, effects of fertiliser type (Smiley and Cook 1973), plant variety, tillage method etc. and 5 - biological factors that influence the host response to disease incidence; predation induced reduction in disease incidence and disease suppressiveness through improvements in plant nutrient availability (Gupta *et al*. 1999, Lemanceau *et al*. 2000).

For the disease suppression known as 'general suppression', the inhibition of pathogenic populations is related to either the activity of the total microflora or diverse microbial-faunal interactions. The 'specific suppression phenomenon' has been attributed to the activity of specific microbial groups (antagonists). Some abiotic factors of soil such as pH and clay content have also been attributed to certain types of disease suppressiveness e.g. fusarium wilts (Lemanceau *et al*. 2000). Disease suppression especially of root diseases is most evident in highly fertile soils.

'General antagonism' (Cook and Baker 1983) has generally been associated with soil fertility. Suppression in such situations is evident as a continuum, the levels being determined by soil fertility and often related to cropping history. Therefore, the interpretation of pathogen assays as indicators of soil biological fertility needs to consider the level of disease suppressiveness of a particular soil and farming system.

8. CONCLUSIONS

Soil-borne plant pathogens affect biological fertility directly and indirectly. Directly, they affect the efficiency of roots to acquire water and nutrients. Indirectly, the reduction in plant biomass resulting from root disease leads to reduced input of the quality and quantity of organic matter that eventually enters the soil. A complex set of organisms, both beneficial and deleterious (including pathogens), is active in soil especially in the rhizosphere region. The environmental conditions, both physical and chemical, determine the balance in their activities that affect soil fertility and plant growth. Management options to enhance soil biological fertility therefore need to consider these interactions and their outcomes both in the bulk and rhizosphere region.

9. ACKNOWLEDGEMENTS

V.V.S.R. Gupta wishes to express his appreciation to his colleagues, in particular Mr. David Roget and Dr. A.D. Rovira at the CSIRO Land and Water in Adelaide for useful and thought provoking discussions regarding the topics related to this chapter. The Grains Research and Development Corporation, Cotton RDC and CSIRO Land and Water support soil biology research in Dr. Gupta's laboratory.

10. REFERENCES

Allen S J 2000 Integrated disease management. How are we doing? *In* Proceedings of the 10th Australian Cotton Conference. pp. 439-442. CRDC. Narrabri, Australia.
Allen S J and Lonergan P A 1998 Trends in diseases in Australian cotton. *In* Proceedings of the 9th Australian Cotton Conference. pp. 541-547. CRDC. Narrabri, Australia.

Antoun H, Beauchamp C J, Goussard N, Chabot R and Lalande R 1998 Potential of *Rhizobium* and *Bradyrhizobium* species as plant growth promoting rhizobacteria on legumes: effect on radishes (*Raphanus sativus* L.). Plant and Soil 204: 57-67.

Baker K F and Cook R J 1974 Biological control of plant pathogens. pp. 433. Freeman and company. San Francisco, USA.

Beare M H, Coleman D C, Crossley D A Jr., Hendrix P F and Odum E P 1995 A hierarchial approach to evaluating the significance of soil biodiversity to biogeochemical cycling. Plant and Soil. 170: 5-22.

Bowen G D and Rovira A D 1999 The rhizosphere and its improvements to plant growth. Advances in Agronomy 66: 1-102.

Cotterill P J and Sivasithamparam K 1989 An autecological study of the take-all fungus (*Gaeumannomyces graminis* var. *tritici*) in Western Australia. Australian Journal of Agricultural Research 40: 229-240.

Chakraborty S and Warcup J H 1985 Reduction of take-all by mycophagous amoebae in pot bioassays. *In* The Ecology and Management of Soil-borne Plant Pathogens. C A Parker, K A Moore, P T W Wong, A D Rovira and J F Kollmorgan (eds.) pp. 14-16. The American Phytopathological Society. St. Paul, Minnesota.

Cook R J and Baker K F 1983 The nature and practice of biological control of plant pathogens. American Phytopathological Society. St Paul, Minnesota.

Curl E A 1988 The role of soil microfauna in plant-disease suppression. CRC Critical Reviews in Plant Sciences 7: 175-196.

Curl E A, Gudauskas R T, Harper J D and Peterson C M 1983 Modification of the rhizosphere fungus flora by mycophagous insects. *In* Abstracts, 3rd International Mycological Congress. pp. 425. Mycological Society of Japan. Tokyo.

Dalal R C 1997 Longterm phosphorus trends in vertisols under continuous cereal cropping. Australian Journal of Soil Research 35: 327-339.

Dalal R C and Chan K Y 2001 Soil organic matter in rainfed cropping systems of the Australian cereal belt. Australian Journal of Soil Research 39: 435-464.

Davies F T Jr, Potter J R and Linderman R G 1992 Mycorrhiza and repeated drought exposure affect drought resistance and extraradical hyphae development of pepper plants independent of plant size and nutrient content. Journal of Plant Physiology 139: 289-294.

Dewan M M and Sivasithamparam K 1990 Differences in pathogenicity of isolates of the take-all fungus from roots of wheat and ryegrass. Soil Biology and Biochemistry 22: 119-122.

Doran J W and Safely M 1997 Defining and assessing soil health and sustainable productivity. *In* Biological Indicators of Soil Health. C E Pankhurst, B Doube and V V S R Gupta (eds.) pp. 1-28. CAB International. Oxon, UK.

Duffy B and Défago G 2000 How pathogens influence biological control. *In* Proceedings of the 5th International PGPR Conference. October-November 2000. Argentina.

Fernando W G D and Linderman R G 1997 The effect of mycorrhizal (*Glomus intraradices*) colonization on the development of root and stem rot (*Phytophthora vignae*) of cowpea. Journal of the National Science Council of Sri Lanka 25: 39-47.

Garrett S D 1970 Pathogenic root-infecting fungi. Cambridge Univ. Press. London.

Gill J S, Sivasithamparam K and Smettem K R 2002 Size of bare-patches in wheat caused by *Rhizoctonia solani* AG-8 is determined by the established mycelial network at sowing. Soil Biology and Biochemistry. 34: 889-893.

Graham J H and Menge J A 1982 Influence of vesicular –arbuscular mycorrhizae and soil phosphate on take-all disease of wheat. Phytopathology 72: 95-98.

Gupta V V S R 1994 The impact of soil and crop management practices on the dynamics of soil microfauna and mesofauna. *In* Soil Biota-Management in Sustainable Farming Systems. C E Pankhurst, B M Doube, VVSR Gupta and P R Grace (eds.) pp. 107-124. CSIRO. Australia.

Gupta V V S R and Germida J J 1988 Distribution of microbial biomass and its activity in different soil aggregate size classes as affected by cultivation. Soil Biology and Biochemistry 20: 777-787.

Gupta V V S R and Yeates G W 1997 Soil microfauna as indicators of soil health. *In* Biological Indicators of Soil Health. C E Pankhurst, B Doube and V V S R Gupta (eds.) pp. 201-233. CAB International. Oxon, UK.

Gupta V V S R and Neate S M 1999 Root disease incidence-A simple phenomenon or a product of diverse microbial/biological interactions. *In* Proceedings of the First Australasian SoilBorne Disease symposium. R C Magarey (ed.) pp. 3-4. BSES. Brisbane, Australia.

Gupta V V S R, Coles R B, Kranz B D and Pankhurst C E 1995 Protozoan - microfloral interactions: effects of microbial metabolites on soil protozoa. Australian Microbiology 16: PO3.18.

Gupta V V S R, Coles R, McClure S, Neate S M, Roget D K and Pankhurst C E 1996 The role of micro-fauna in the survival and effectiveness of plant pathogenic fungi in soil. ASM & NZSM Conference. Christchurch. Australian Microbiology 17: A64.

Gupta V V S R, Neate S M and Dumitrescu I 1999. Effects of microfauna and mesofauna on *Rhizoctonia solani* in a South Australian soil. *In* Proceedings of the First Australasian soil-borne disease symposium. R.C. Magarey (ed.) pp. 134-136. BSES. Brisbane, Australia.

Hendrix P F, Parmelee R W, Crossley D A, Coleman D C, Odum E P and Groffman P M 1986 Detritus food webs in conventional and no-tillage agroecosystems. Bioscience 36: 374-80.

Herdina and Roget D K 2000 Prediction of take-all disease risk in field soils using a rapid and quantitative DNA soil assay. Plant and Soil 227: 87-98.

Herdina, Roget D K, Coppi J A and Gupta V V S R 2001 Investigation of the mechanisms of disease suppression in soil using DNA techniques. *In* Proceedings of the Second Australasian Soil-borne Disease Symposium. I J Porter *et al.* (eds.) pp. 88-89. Department of Natural Resources and Environment. Victoria, Australia.

Hoper H and Alabouvette C 1996 Importance of physical and chemical soil properties in the suppressiveness of soils to plant diseases. European Journal of Soil Science 32: 41-58.

Hornby D and Bateman G L 1997 Potential use of plant pathogens as bioindicators of soil health. *In* Biological Indicators of Soil Health. C.E. Pankhurst, B.Doube and V V S R Gupta (eds.) pp. 179-200. CAB International. Oxon, UK.

Huber D M 1981 The role of nutrients and chemicals. *In* Biology and control of take-all. M J C Asher and P J Shipton (eds.) pp. 317- 341. Academic Press. London.

Kloepper J W, Rodriguez-Kabana R, Zehnder G W, Murphy J F, Sikora E and Fernandez C 1999 Plant root-bacterial interactions in biological control of soilborne diseases and potential extension to systemic and foliar diseases. Australasian Plant Pathology 28: 21-26.

Kloepper J W, Reddy M S, Rodríguez-Kábana R, Kenney D S, Martinez-Ochoa N, Kokalis-Burelle N and Arthur K 2000 Development of an integrated biological preparation for growth enhancement of various vegetable transplant plugs suppressive to plant diseases. *In* Proceedings of the 5th International PGPR conference, October-November 2000. Argentina.

Kobayashi D and Palumbo J D 2000 Bacterial endophytes and their effects on plants and uses in agriculture. *In* Microbial endophytes. C W Bacon and J F White, (eds) pp. 199-233. Marcel Dekker Inc. New York.

Lartey R T, Curl E A, Peterson C M and Harper J D 1989 Mycophagous grazing and food preference of *Proisotoma minuta* (Collembola: Isotomidae) and *Onychiurus encarpatus* (Collembola: Onychiuridae). Environmental Entomology 18: 334-337.

Lee K E and Foster R C 1991 Soil fauna and soil structure. Australian Journal of Soil Research 29: 745-775.

Lemanceau Ph, Steinberg C, Thomas D J I, Edel V, Raaijmakers J M and Alabouvette C 2000 Natural suppressiveness to soil borne diseases. *In* Proceedings of the 5th International PGPR conference, October-November 2000. Argentina.

Lussenhop J and Wicklow D T 1984 Invertebrates disperse fungal propagules in soil. Bulletin of the Ecological Society 65: 1073-1079.

MacDonald H J and Rovira A D 1985 Development of inoculation technique for *Rhizoctonia solani* and its application to screening cereal cultivars for resistance. *In* Ecology and Management of Soilborne Plant diseases. C A Parker, A D Rovira, K J Moore, P T Wong and J F Kollmorgen (eds.) pp. 174-176. American Phytopathological Society. St Paul, MN, USA.

Mazzola M and Cook R J 1991 Effects of fungal root pathogens on the population dynamics of biocontrol strains of fluorescent pseudomonads in the wheat rhizosphere. Applied and Environmental Microbiology 57: 2171-2178.

Mele P M and Carter M R 1993 Effect of climatic factors on the use of microbial biomass as an indicator of changes in soil organic matter. *In* Soil Organic Matter Dynamics and Sustainability of Tropical Agriculture. K Mulongoy and R Merckx (eds.) pp. 57-63. John Wiley and Sons Ltd. Chichester, UK.

Neate S N 1994 Soil and crop management practices that affect root diseases of crop plants. *In* Soil Biota-Management in Sustainable Farming Systems. C E Pankhurst, B M Doube, V V S R Gupta and P R Grace (eds.) pp. 96-106. CSIRO. Australia.

Nilsson H E 1969 Studies on root and foot rot diseases of cereals and grasses. I. On resistance of *Ophiobolus graminis* Sacc. Lanbruks. Annaler 35: 275-807.

Old K M 1986 Mycophagous amoebae: Their biology and significance in the ecology of soil-borne plant pathogens. Progress in Protozoology 1: 163-194.

Pankhurst C E, Hawke B G, McDonald H J, Kirkby C A, Buckerfield J C, Michelsen P, Gupta V V S R and Doube B M 1995 Evaluation of soil biological properties as potential bioindicators of soil health. Australian Journal of Experimental Agriculture 35. 1015-1028.

Roget D K 1995 Decline in root rot (*Rhizoctonia solani* AG-8) in wheat in a tillage and rotation experiment at Avon, South Australia. Australian Journal of Experimental Agriculture 35: 1009-1013.

Roget D K 2001 Prediction modelling of soilborne plant diseases. Australasian Plant Pathology 30: 85-89.

Roget D K, Neate S N and Rovira A D 1996 The effect of sowing point design and tillage practices on the incidence of rhizoctonia root rot, take-all and cereal cyst nematode. Australian Journal of Experimental Agriculture. 36: 683-693.

Roget D K, Coppi J A, Herdina and Gupta V V S R 1999 Assessment of suppression to *Rhizoctonia solani* in a range of soils across SE Australia. *In* Proceedings of the First Australasian Soil-borne Disease Symposium. R C Magarey (ed.) pp. 129-130. BSES. Brisbane, Australia.

Roper M M and Gupta V V S R 1995. Management practices and soil biota. Australian Journal of Soil Research 33: 321-39.

Rovira A D 1986 Influence of crop rotation and tillage on Rhizoctonia bare patch of wheat. Phytopathology 76: 669-673.

Rovira A D, Elliott L F and Cook R J 1990 The impact of cropping systems on rhizosphere organisms affecting plant health. *In* The Rhizosphere. J M Lynch (ed.) pp. 389-436. Wiley Interscience. Chichester, UK.

Shaw C G and Kile G A 1991 Armillaria Root Disease. Agriculture Handbook No 691. USDA. Washington.

Simon A and Sivasithamparam K 1989 Pathogen suppression: A case study in biological suppression of *Gaeumannomyces graminis* var. *tritici* in soil. Soil Biology and Biochemistry 21: 331-337.

Sivasithamparam K 1993 Ecology of root-infecting pathogenic fungi in mediterranean environments. Advances in Plant Pathology 10: 245-279.

Sivasithamparam K 1998 Root cortex-the final frontier for the biocontrol of root-rot with fungal antagonists: a case study on a sterile red fungus. Annual Review of Phytopathology 36: 439-452.

Smiley R W and Cook R J 1973 Relationship between take-all of wheat and rhizosphere pH in soils fertilised with ammonium and nitrate nitrogen. Phytopathology 63: 882-890.

Smith S E, Read D J 1997 Mycorrhizal symbiosis. Academic Press. London. UK.

Sturtz A V, Christie B R and Nowak J 2000 Bacterial endophytes: Potential role in developing sustainable systems of crop production.   Critical Reviews in Plant Sciences 19: 1-30.

Subramanian K S and Charest S 1999 Acquisition of N by external hyphae of an arbuscular mycorrhizal fungus and its impact on physiological responses in maize under drought-stressed and well-watered conditions. Mycorrhiza 9: 69-75.

Thongbai P, Hannam R J, Graham R D and Webb M J 1993 Interaction between zinc nutritional status of cereals and Rhizoctonia root rot severity. I. Field observations. Plant and Soil 153: 207-214.

Tisdall J M 1991 Fungal hyphae and the structural stability of soil. Australian Journal of Soil Research 29: 729-743.

Tisdall J M, Smith S E and Rengasamy P 1997 Aggregation of soil by fungal hyphae. Australian Journal of Soil Research 35: 55-60.

Tu J C 1978 Protection of soybean from severe phytophthora root rot by *Rhizobium*. Physiological Plant Pathology 12: 233.

van Veen J A and Heijnen C E 1994 The fate and activity of microorganisms introduced in soil. *In* Soil Biota-Management in Sustainable Farming Systems. C E Pankhurst, B M Doube, V V S R Gupta, P R Grace (eds.) pp. 107-124. CSIRO. Australia.

Van Loon L C, Bakker P A H M and Poeterse C M J 1998 Systemic resistance induced by rhizosphere bacteria. Annual Review of Phytopathology 36: 453-483.

Wilhelm N S, Graham R D and Rovira A D 1990 Control of Mn status and infection rate by genotype of both host and pathogen in the wheat take-all interaction. Plant and Soil 123: 267-275.

Wiseman B M, Neate S M, Ophel Keller K and Smith S E 1996. Suppression of *Rhizoctonia solani* anastomosis group 8 in Australia and its biological nature. Soil Biology and Biochemistry 28: 727-732.

Chapter 9

Relevance of Interactions amongst Soil Microorganisms to Soil Biological Fertility

Wendy M. Williamson[1] and David A. Wardle[2]

[1] *Landcare Research, PO Box 69, Lincoln, New Zealand*

[2] *Department of Animal & Plant Sciences, University of Sheffield, Sheffield S10 2TN, U.K.*

1. INTRODUCTION

In this chapter we discuss trophic interactions within the soil food web, with emphasis on soil microorganisms, and how these interactions influence the release of plant-available nutrients. The theme is that soil fertility is causally linked to interactions between soil biotic components and is exemplified by the interdependence and non-random location of plants and soil organisms. We use a conceptual model of detritus decomposition divided into bacterial and fungal energy channels to illustrate how non-random interactions, such as substrate quality and the presence of plant roots, influence soil fertility. Finally, an ecological perspective summarises land management effects on nutrient cycling and identifies key mechanisms that control the fertility of soil in managed systems.

L.K. Abbott & D.V. Murphy (eds) Soil Biological Fertility - A Key to Sustainable Land Use in Agriculture. 187-201.
© 2007 *Springer.*

2.   NON-RANDOM LOCATION OF SOIL ORGANIC MATTER

Organic matter is not randomly located within soil. For example, sulfur bacteria can align themselves along the crystal planes of sulfur-containing clay minerals (Edwards *et al.* 2000).   Such specific biotic and abiotic interactions will act as primary foci for the accumulation of soil organic matter, the build up of which will provide microbial habitats, promote microbial activity and spatially segregate processes that soil microorganisms mediate. The associations of soil microorganisms in specific microhabitats within the soil matrix can be deduced from the locations of ATP and enzyme activity. Ladd (1972) found ATP and active enzymes located with the fine clay fraction and in fragments of plant material (Figure 1), suggesting that clay minerals and organic matter serve as hotspots for biological activity in soil. It is also of interest that as water-stable soil particles become smaller, the C-to-N ratios decrease. This is indicative of plant residues (which are characterised by higher C-to-N ratios) disappearing from the soil organic matter and being replaced by microbial biomass (Ahmed and Oades 1984). Chemotactic foraging behaviour of plant roots and soil microbes is dependent on this non-random location of organic matter hotspots for the establishment of chemical gradients that induce plant and microbial responses of attraction or repulsion (Ahmed and Oades 1984, Tinker 1984). Spatially distinct patches of organic matter enhances nutrient uptake by plants to a greater extent than when the organic matter is homogeneously distributed through the soil, with enhanced root growth localised within patches (Bonkowski *et al.* 2000).

While many soil microbes are adapted for rapid growth in the presence of plant roots (see Chapter 5 this volume), a large number survive outside the rhizosphere. Organisms that live outside the rhizosphere inhabit a much harsher environment and the patches of organic matter they occupy comprise less than 5% of the soil volume (Griffiths 1994). These zones are characterised by low microbial metabolic activity, with spores, cysts and dormancy dominating the physiological and morphological status of the community (Tinker 1984). However, the microorganisms that are active in this zone are likely to determine the direction that plant roots will follow, since the active microbes will utilise resources spatially removed from the root, establishing concentration gradients to which plants and soil flora and fauna respond. Free-living organisms tend to be dominated by K-selected species (which are characterised by low metabolic activity) and while they may have specific enzymatic capabilities to fix N or utilise complex substrates, competition for C probably limits their effectiveness (Tinker 1984). Therefore, away from the plant roots, plant-available nutrients are bound up in complex organic matter, and the soil environment is effectively a nutrient-poor matrix dotted with discrete islands rich in organic matter and biotic activity that are connected by chemical gradient bridges.

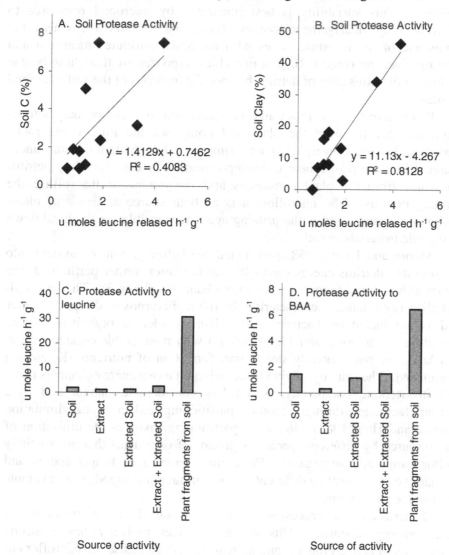

Figure 1 Activity and location of soil enzymes as a response to presence of plant litter in the soil. (A,B) Correlation of soil protease activity (expressed as μmoles leucine released from Z-phenylalanyl leucine per hour per g soil dry weight) with (A) organic C content of the soil (%) and (B) clay content of soil (%). (C, D) The protease activity towards either (C) Z-phenylalanyl leucine or (D) benzoyl arginine amide (BAA), expressed as μmoles leucine released per hour per g soil dry weight, from specific fraction of soil. Derived from (Ladd 1972).

3.   ENERGY CHANNELS AND SOIL FOOD WEBS

The magnitude of microbial response to soil fertility tends to be variable. This variability is best illustrated by microbial responses to different crop management systems (perennial vs annual), where the magnitude of the response varies with microbial attributes measured and crop management (Figure 2). The microbial responses are thought to be due to differential utilisation of detritus by specific fractions of the soil microbial biomass.

Plants give rise to primary and secondary detritus. The primary detritus originates directly from the plant and comprises the litter (dead roots, branches, leaves, flowers etc.) and simple carbon-containing substances secreted into the rhizosphere (rhizodeposition), whereas secondary detritus originates from the plant consumers and the organisms that utilise the primary detritus. The microflora acts as both source and sink of plant-available nutrients, and is the priming agent responsible for the breakdown of organic materials in soil.

Moore and Hunt (1988) partitioned the below-ground ecosystem into the root and detritus energy channels, and the latter further partitioned into bacterial-based and fungal-based energy channels (Figure 3). The bacterial-based energy channel is characterised by the rapid turnover of organic matter and is dominated by bacterial metabolism (r-selected organisms). The fungal-based energy channel is associated with more stable organic matter, and has slow but relatively steady transformation of nutrients (K-selected organisms). The primary driving force behind the two energy channels is the soil invertebrate community, which has considerable habitat and food overlap facilitated through resource partitioning and body size limitation (Moore and Hunt 1988). Resource partitioning involves the utilisation of one resource by different species or groups of organisms that are spatially and/or temporally segregated. Preferential grazing of fungal spores and hyphae (young vs old) by different species of microarthropods is an example of resource partitioning.

Bacteria and the processes that they mediate dominate the bacterial-based energy channel. This energy channel typically has a strong association of active protozoa and nematodes, which utilise the microflora as a resource and are dependent on available water for motility. This interaction between the microfauna and microflora of the bacterial-based energy channel is highly relevant to the sustained biological fertility of soil. For example, protozoa are effective at releasing significant quantities of mineral N into the soil environment, especially in the rhizosphere; a 5% increase in protozoa numbers led to a 21% increase in mineral N, while a

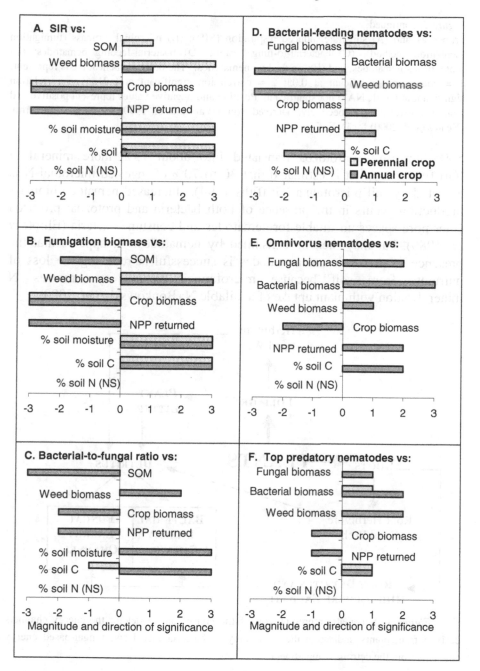

Figure 2 Levels of significance of correlation between below-ground properties in two cropping systems (one annual: maize (Zea mays L.) and one perennial: asparagus (Asparagus officinale L)) bars within the panels represent the strength between a range of belowground

Figure 2 continued on page 192

Figure 2, continued

properties and (A) substrate induced respiration (SIR); (B) microbial biomass (fumigation incubation method); (C) bacterial-to-fungal ratio; (D) bacterial-feeding nematodes; (E) omnivorous nematodes; and (F) top feeding nematodes. The numbers beneath axis represent: 1 = correlation significant at 0.05; 2 = correlation significant at 0.01; 3 = correlation significant at 0.001, NA = not significant. Positive and negative values represent positive and negative correlations respectively. Derived from (Wardle et al. 1999) for A, B, C and from (Yeates et al. 2000) for D, E, F.

50% increase in protozoa translated into about 73% more mineral N (Griffiths 1994). It is estimated that 30 to 70% of ingested microbial-N is excreted by soil microfauna (Griffiths 1994). Increased nematode biomass production occurs in the presence of both bacteria and protozoa; protozoa enter pore spaces unsuitable for nematodes and consume bacteria (Elliott *et al.* 1980), but may then be consumed by nematodes. It appears that the presence of protozoa and nematodes is successful at reducing the loss of nutrients from soil because microfaunal grazing synchronises N mineralisation with plant uptake of available N (Bonkowski *et al.* 2000).

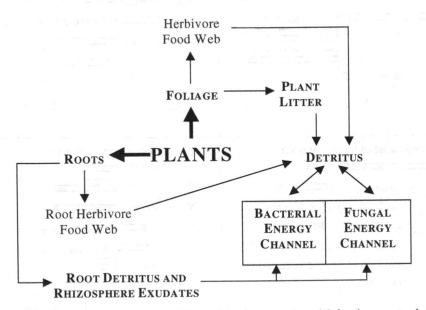

Figure 3 Schematic representation of the conceptual model for the energy channels derived from plants, indicating the dichotomy of bacterial-based and fungal-based energy channels from the detritus compartment.

Grazing by protozoa directly affects the morphological characteristics of the soil bacterial community. Medium-sized bacterial cells dominate the population when protozoan grazing is low. However, as the grazing pressure increases, situations and only small and large bacteria resit grazing (Hahn and Hofle 1999). The protozoan grazing effect is due to smaller sized bacteria being able to occupy pore spaces that are too small for protozoa to enter or that have a pore neck size that limits protozoan access. Alternatively, small bacteria may aggregate into multicellular communities that have an effective size unsuitable for protozoan ingestion (Shapiro 1995). Some large bacteria may escape initial grazing pressure and as a result of reduced competition rapidly grow too large for subsequent protozoan grazing (Hahn and Hofle 1999).

With regard to the fungal-based energy channel, fungal-invertebrate interactions are typified by microarthropod grazing on fungal mycelium and spores. Selective grazing by microarthropods can alter fungal community structure, and therefore strongly influence the dynamics of litter decomposition through feeding interactions that preferentially promote intraspecific and interspecific fungal activity (Newell 1984). The fungal-based energy channel is usually favoured by low disturbance (e.g. no-till systems), with more nutrients immobilised in the fungal biomass than when bacterial-based energy channels are favoured.

4. INTERACTIONS OF THE SOIL FOOD WEB

Microbial growth is promoted when interactions between the soil microflora and soil invertebrates. The soil invertebrate community has three functional compartments, micropredators, litter transformers and ecosystem engineers (Lavelle *et al.* 1995), which operate within and between trophic levels by a network of interactions. Micropredators are the small invertebrates (protozoa, nematodes and microarthropods) whose activity is dependent on the body size of their prey or nutrition gained from them. Prey size tends to determine the detritus energy channels with which the micropredators are principally associated.

Litter transformers have two direct effects on the soil ecosystem. Firstly, they are responsible for comminutation of litter. This disrupts cellular integrity and greatly increases the surface area of the resource available for microbial attack by mechanical and enzymatic means. Secondly, the litter transformers' faecal pellets represent dynamic sites of concentrated organic matter that are strongly associated with microbial activity and are collectively referred to as the external rumen (Lavelle *et al.* 1995).

Ecosystem engineers, such as earthworms and termites, build more permanent organo-mineral structures. These structures provide additional habitats for soil fauna and microorganisms. The ecosystem engineers develop strong mutualistic interactions with the microflora in their gut cavities, allowing macrofauna with low digestive capabilities to consume food that would otherwise be unsuitable. The gut environment is essentially one of an unlimited aqueous phase, with temperature, moisture and pH conditions conducive for microbial metabolism and reproduction, and greatly favours bacterial-based energy channel processes (Daniel and Anderson 1992).

5.   RESOURCES FOR THE SOIL FOOD WEB

Resource (residue) quality alters during decomposition. Decomposition is affected by the biotic and abiotic factors that influence microbial activity and associated interactions (Wardle 2001). Therefore, even though organic matter is not randomly located in soil, the direction and magnitude of biotic responses to substrate quality and soil fertility are often difficult to predict (Figure 2).

The addition of plant and animal residues to soil tends to increase soil enzyme activity by two mechanisms. First, enhanced microbial activity in response to residue addition is typified by increased activity of the bacterial-based energy channel. Second, residue-associated enzymes, some of which retain activity for a considerable period of time, enhance decomposition by reducing the direct energy expenditure by the microflora (Dick and Tabatabai 1993). In addition to soil physical factors that affect enzyme activity (such as pH, temperature and presence of inhibitory compounds), vegetation type and vegetation quality are principal determinants of soil enzyme activity (Dick and Tabatabai 1993).

The rhizosphere is a principal functional unit in regulating biological soil fertility, and is typified by increased bacterial activity due to more readily available C from roots and root exudates. Rhizosphere-associated bacteria are more active than those inhabiting non-rhizosphere zones (Lynch and Whipps 1990) and tend be larger (Foster 1988). Further, the doubling time of bacteria in the rhizosphere is less than that in non-rhizosphere soil, and this may be due to higher grazing intensities in the rhizosphere that keep the bacteria in logarithmic growth. For example, *Pseudomonas* spp. can have a generation time of five to six hours near the root tip compared with 77 hours in bulk soil. In contrast, the generation time of *Bacillus* spp. may be 39 hours in the rhizosphere compared with 100 hours when not associated with roots (Anderson 1988). The enhanced microbial activity close to the root is associated with greater protozoan and nematode activity and the microbial and microfaunal populations tend to follow the spatial and temporal root tip growth. There is a reduction in microbial biomass that is correlated with the distance from plant roots. Similarly, the microbial biomass associated with older roots is lower than that of younger more active roots (Newman 1985). These spatial distributions are consistent with the non-random location of microbes in soil.

In the non-rhizosphere soil, fungi dominate the microbial biomass, suggesting that very different interactions occur away from plant roots. The translocation of resources between regions of the same fungal colony, and even between fungi, is considered central to the generalist ability of saprotrophic fungi (Thrower and Thrower 1961, Connolly *et al*. 1999). The ability to translocate nutrients is not a universal feature of fungi, nor is it expressed in an unregulated manner (Thrower and Thrower 1961). While some fungi can only translocate when they encounter nutritionally rich substrates, most can move C from older to younger tissue (Thrower and Thrower 1961). Although it is generally assumed that the fungi translocate nutrients only from decomposing organic matter, Connolly *et al*. (1999) showed that some fungi could extract and subsequently translocate nutrients directly from the mineral horizon to the surface organic matter, which may represent a very important mechanism for spatial dynamics and redistribution of nutrients in the soil. Net-N immobilisation in surface plant residues is directly associated with fungal activity (Frey *et al*. 2000) and reflects the lower turnover of available resource when processes are dominated by fungal-based energy channel processes (i.e. grazed by microarthropods rather than protozoa). This indicates that any mineral N produced in surface litter is initially unavailable to plants and suggests that this pool of N if subsequently translocated may permit fungi to exploit resources of less favourable C-to-N ratios (Frey *et al*. 2000).

Earthworm mucus can be likened to rhizodeposition by plants and gut-associated processes have parallels with those driven by root-elongation (Martin *et al*. 1987). The passage of microbes through the earthworm gut usually alters microbial community structure (Edwards and Fletcher 1988, Daniel and Anderson 1992). The feeding and burrowing activity of earthworms introduces fungi, bacteria and litter fragments to different sites within the soil resulting in temporal and spatial dynamics that the microbes could not have achieved independently (Tiunov and Scheu 2000). In addition, earthworm casts and the faecal pellets of arthropods can be more compact than the surrounding soil (Chauvel *et al*. 1999), which favours bacterial processes over fungal due to the available pore size being smaller. However, fungi rapidly colonise the outside of the pellets and the external rumen process of organic matter breakdown continues (Swift *et al*. 1979).

Competition for soil nutrients means that many organisms increase their competitiveness by inhibiting the growth of other microbes directly through the production of antimicrobial agents (Cain *et al*. 2000). Other mechanisms by which negative interactions between soil microorganisms may work include the production of extracellular agents capable of causing cell lysis (Kope and Fortin 1990), reduction in extracellular polysaccharides (Barrion and Habte 1988) and sequestering of ferric ions, such as action of siderophores (Fogel 1985). Production of extracellular polysaccharides, which help microorganisms resist adverse environmental conditions, have been shown to be substantially reduced through the competitive effects of soil actinomycetes (Barrion and Habte 1988).

6. A PERSPECTIVE OF SOIL BIOLOGICAL FERTILITY

Agricultural intensification, which generally involves the increasing use of mechanical and synthetic inputs and the deliberate reduction of aboveground biodiversity, has important direct and indirect effects on soil function. Decreasing levels of soil organic matter, which may reduce the soil microbial biomass and therefore the biological fertility of the soil, can adversely affect belowground nutrient cycling processes. Although soil organic matter stocks directly determine the fertility of the soil (Tiessen *et al.* 1994), the initial status of a soil's nutrient reserves do not necessarily indicate how responsive a soil will be to cropping pressure. It is the rate at which standing organic matter is lost, rather than the absolute amount of organic matter, that is the key determinant of long-term soil fertility. Soils with higher nutrient contents initially tend to lose a larger proportion of their organic matter than soils of lower status (i.e. a non-linear response), when agricultural intensification occurs (Sparling *et al.* 2000). The constraints on soil microbial activity do not necessarily determine the absolute state of the habitat, but rather, the degree and direction of change in the physical, chemical and biological properties dictates the microbial response to soil management, such as cropping and tillage. In general, increasing cropping pressure or tillage decreases soil organic resources (Davidson and Ackerman 1993) and consequently decreases the metabolic activity of soil microflora and microfauna (Sparling *et al.* 2000).

Agricultural practices that promote soil disturbance, such as tillage, have important effects on soil food web structure, and tend to adversely affect larger-bodied organisms (e.g. earthworms, arthropods) more than smaller animals such as protozoa and nematodes (Wardle 1995). Therefore, soil disturbance by tillage tends to promote the bacterial-based energy channel at the expense of the fungal-based channel and tends to disadvantage larger animals, reducing flow-on benefits from gut-associated mutualistic microbial interactions (Hendrix *et al.* 1986).   However, cultivation also generates buried resource islands, which may contribute to greater spatial heterogeneity in soil (Wardle *et al.* 1999).

In low intensity agricultural systems herbicides are frequently used in the place of tillage to control weeds. Indirect effects of herbicides have been postulated as one of the reasons for the differential responses of soil food-web components to no-tillage agriculture (Hendrix and Parmelee 1985). Firstly, the herbicides serve as C substrates for bacteria and promote the bacterial-based energy channel, although these effects are generally likely to be minor (Domsch *et al*. 1983). A more important mechanism of altering the below-ground food web involves the changes in plant community structure brought about by herbicide use. The presence of weeds at moderate levels (but at levels below which are likely to interfere with crop production), results in patches of higher quality litter in the soil (Wardle *et al*. 1995); this in turn has important positive effects on components of the soil biota.

Cropping affects soil fertility partly through altering soil biotic interactions. Such alterations can lead to complex interactions in which only some groups appear to be stimulated. However, other groups may have been stimulated, but are subsequently regulated by higher trophic levels, giving an apparent decreased or neutral response to cropping. For example, in a cropping system Wardle *et al*. (1999) found that increased residue additions (sawdust) to a perennial crop promoted fungal, but not bacterial growth. In this case, resource addition promoted top predatory nematodes, but predation pressure induced by these nematodes resulted in regulation at lower trophic levels, preventing bacteria from showing a positive response towards resource addition even though their turnover rate increased. This indicates the importance of studying several trophic levels of the soil food web when investigating the dynamics of specific soil management practices.

The presence of grazing animals can exert important effects upon the decomposer subsystem as well as on nutrient mineralisation processes carried out by the soil biota (Bardgett *et al*. 1998). These include the return of dung and urine, alteration of plant productivity, allocation of C and nutrients to roots (Seastedt 1985), and the quantity of organic matter returned by the plant (Bardgett *et al*. 1998). Continuous grazing of grasses appears to favour the bacterial-based energy channel over fungal in the short-term (Mawdsley and Bardgett 1997). In the longer-term, the net effect of defoliation may be towards the fungal-based energy channel as rhizodeposition declines and the soil microbial community adapts accordingly, with the long-term effect of defoliation tending to diminish microbial activity (Mikola *et al*. 2001). The difference in the nature of resource input between crop production and livestock-based agricultural systems can also strongly influence the nature of the soil food web and soil fertility. Because grazing systems tend to be based on perennial plant species, they are generally associated with soils with higher levels of organic matter, higher levels of soil microbial biomass, and greater populations of the associated soil animals (Srivastava and Singh 1991). However, grazed grasslands can also support lower populations of smaller soil organisms, relative to cropping systems, such as nematodes, which typify the bacterial-based energy channel (Yeates *et al.* 2000).

In summary, soil management practices that promote the maintenance of organic matter and an active microbial biomass, which may be approached through the proper management of organic residues, can maintain crop production through biological mechanisms regulating the supply of plant-available nutrients. The factors that control soil biological fertility are therefore those that regulate the activity of the soil microbial biomass and the interactions with soil fauna. Decomposition processes and nutrient cycling occur in a defined manner, however the dynamics within different systems will alter the multitude of possible interactions due to differences in substrate quality. Therefore it is difficult to make specific predictions of biological response to different soil management practices. This difficulty is despite the fact that soil processes are initiated and maintained by non-random processes both spatially and temporally.

7.   ACKNOWLEDGEMENTS

Financial support was provided by AGMARDT (The Agricultural and Marketing Research and Development Trust) to W. Williamson.

8.   REFERENCES

Ahmed M and Oades J M 1984 Distribution of organic matter and adenosine triphosphate after fractionation of soils by physical procedures. Soil Biology and Biochemistry 16: 465-470.

Anderson J M 1988 Spatiotemperal effects of invertebrates on soil processes. Biology and Fertility of Soils 6: 216-227.

Bardgett R D, Wardle D A and Yeates G W 1998 Linking aboveground and belowground interactions: how plant responses to foliar herbivory influence soil organisms. Soil Biology and Biochemistry 30: 1867-1878.

Barrion M and Habte M 1988 Interaction of *Bradyrhizobium* sp with an antagonistic actinomycete in culture medium and in soil. Biology and Fertility of Soils 6: 306-310.

Bonkowski M, Griffiths B and Scrimgeour C 2000 Substrate heterogeneity and microfauna in soil organic 'hotspots' as determinants of nitrogen capture and growth of ryegrass. Applied Soil Ecology 14: 37-53.

Cain C C, Henry A T, Waldo R H, Casida L J and Falkinham J O 2000 Identification and characteristics of a novel *Burkholderia* strain with broad-spectrum antimicrobial activity. Applied and Environmental Microbiology 66: 4139-4141.

Chauvel A, Grimaldi M, Barros E, Blanchart E, Desjardins T, Sarrazin M and Lavelle P 1999 Pasture damage by an Amazonian earthworm. Nature 398: 32-33.

Connolly J H, Shortle W C and Jellison J 1999 Translocation and incorporation of strontium carbonate derived strontium into calcium oxalate crystals by the wood decay fungus *Resinicium bicolor*. Canadian Journal of Botany 77: 179-187.

Daniel O and Anderson J M 1992 Microbial biomass and activity in contrasting soil materials after passage through the gut of the earthworm *Lumbricus rubellus* Hoffmaster. Soil Biology and Biochemistry 24: 465-470.

Davidson E A and Ackerman I L 1993 Changes in soil carbon inventories following cultivation of previously untilled soils. Biogeochemistry 20: 161-193.

Dick W A and Tabatabai M A 1993 Significance and potential uses of soil enzymes. *In* Soil Microbial Ecology: Applications in Agriculture and Environmental Management. F B Metting (ed.) pp. 95-127. Marcel Dekker Inc. New York.

Domsch K H, Jagnow G and Anderson T H 1983 An ecological concept for the assessment of side-effects of agrochemicals on soil microorganisms. Research Reviews 86: 65-105.

Edwards C A and Fletcher K E 1988 Interactions between earthworms and microorganisms in organic matter breakdown. Agriculture, Ecosystems and Environment 24: 235-247.

Edwards K J, Bond P L and Banfield J F 2000 Characteristics of attachment and growth of *Thiobacillus caldus* on sulphide minerals: a chemotactic response to sulphur minerals. Environment Microbiology 2: 324-332.

Elliott E T, Anderson R V, Coleman D C and Cole C V 1980 Habitable pore space and microbial trophic interactions. Oikos 35: 327-335.

Fogel R 1985 Roots as primary producers in belowground ecosystems. *In:* Ecological Interactions in Soil. A H Fritter (ed.) pp. 23-36. Blackwell Scientific Publications. Oxford.

Foster R C 1988 Microenvironments of soil microorganisms. Biology and Fertility of Soils 6: 189-203.

Frey S D, Elliott E T, Paustian K and Peterson G A 2000 Fungal translocation as a mechanism for soil nitrogen inputs to surface residue decomposition in a no-tillage agroecosystem. Soil Biology and Biochemistry 32: 689-698.

Griffiths B S 1994 Soil nutrient flow. *In* Soil Protozoa. J F Darbyshire (ed.) pp. 65-91. CAB International. Wallingford, UK.

Hahn M W and Hofle M G 1999 Flagellate predation on a bacterial model community: interplay of size selective grazing, specific bacterial cell size and bacterial community composition. Applied and Environmental Microbiology 65: 4863-4872.

Hendrix P F and Parmelee R W 1985 Decomposition, nutrient loss and microarthropod densities in herbicide-treated grass litter in a georgia piedmont agroecosystem. Soil Biology and Biochemistry 17: 421-428.

Hendrix P F, Parmelee R W, Crossley D A, Coleman D C, Odum E P and Groffman P M 1986 Detritus food webs in conventional and no tillage agroecosystems. Bioscience 36: 374-380.

Kope H H and Fortin J A 1990 Antifungal activity in culture filtrates of the ectomycorrhizal fungus *Pisolithus tinctorius*. Canadian Journal of Botany 68: 1254-1259.

Ladd J N 1972 Properties of proteolytic enzymes extracted from soil. Soil Biology and Biochemistry 4: 227-237.

Lavelle P, Lattaud C, Trigo D and Barois I 1995 Mutualism and biodiversity in soils. *In* The Signifiance and Regulation of Soil Biodiversity. H P Collins, G P Robertsono and M J Klug (eds.) pp. 23-33. Kluwer Academic Publishers. The Netherlands.

Lynch J M and Whipps J M 1990 Substrate flow in the rhizosphere. Plant and Soil 129: 1-10.

Martin A, Cortez J, Barois I and Lavelle P 1987 Les mucus intestinaux de ver de terre, moteur de leurs interactions avec la micrflore. Rev. Ecol. Biol. Sol. 24: 549-558.

Mawdsley J L and Bardgett R D 1997 Continuous defoliation of perennial ryegrass (*Lolium perenne*) and white clover (*Trifolium repens*) and associated changes in the composition and activity of the microbial population of an upland grassland soil. Biology and Fertility of Soils 24: 52-58.

Mikola J, Yeates G W, Barker G M, Wardle D A and Bonner K I 2001 Effects of defoliation intensity on soil food-web properties in an experimental grassland community. Oikos 92: 333-343.

Moore J C and Hunt H W 1988 Resource compartmentation and the stability of real ecosystems. Nature 333: 261-263.

Newell K 1984 Interaction between two decomposer basidiomycetes and a collembolan under sitka spruce: distribution, abundance and selective grazing. Soil Biology and Biochemistry 16: 227-233.

Newman E I 1985 The rhizosphere: carbon sources and microbial populations. *In* Ecological Interactions in Soil. A H Fritter (ed.) pp. 107-121. Blackwell Scientific Publications. Oxford.

Seastedt T R 1985 Maximisation of primary and secondary productivity by grazers. American Naturalist 126: 559-564.

Shapiro J A 1995 The significances of bacterial colony patterns. BioEssaya 17: 597-607.

Sparling G P, Schipper L A, Hewitt A E and Degens B P 2000 Resistance to cropping pressure of two New Zealand soils with contrasting mineralogy. Australian Journal of Soil Research 38: 85-100.

Srivastava S C and Singh J S 1991 Microbial C, N and P in dry tropical forest soils; effects of alternative land uses and nutrient flux. Soil Biology and Biochemistry 23: 117-124.

Swift M J, Heal O W and Anderson J M 1979 Decomposition in Terrestrial Ecosystems. Blackwell Scientific Publishers. Great Britain.

Thrower S L and Thrower L B 1961 Transport of carbon in fungal mycelium. Nature 190: 823-824.

Tiessen H, Cuevas E and Chacon P 1994 The role of soil organic matter in sustaining soil fertility. Nature 371: 783-785.

Tinker P B 1984 The role of microorganisms in mediating and facilitating the uptake of plant nutrients from soil. Plant and Soil 76: 77-91.

Tiunov A V and Scheu S 2000 Microfungal communities in soil, litter and casts of *Lumbricus terrestris* L. (Lumbricidae): a laboratory experiment. Applied Soil Ecology 14: 17-26.

Wardle D A 1995 Impacts of disturbance on detritus food webs in agro-ecosystems of contrasting tillage and weed management practices. *In* Advances in Ecological Research. M Begon and A H Fitter (eds.) pp. 105-185. Academic Press. London.

Wardle D A 2001 Communities and Ecosystems: Linking the Aboveground and Belowground Components. Princeton University Press. Princeton, USA.

Wardle D A, Yeates G W, Nicholson K S, Bonner K I and Watson R N 1999 Response of soil microbial biomass dynamics, activity and plant litter decomposition to agriculture intensification over a seven year period. Soil Biology and Biochemistry 31: 1707-1720.

Wardle D A, Yeates G W, Watson R N and Nicholson K S 1995 The detritus food-web and the diversity of soil fauna as indicators of disturbance regimes in agro-ecosystems. Plant and Soil 170: 35-43.

Yeates G W, Hawke M F and Rijkse W C 2000 Changes in soil fauna and soil conditions under *Pinus radiata* agroforestry regimes during a 25-year tree rotation. Biology and Fertility of Soils 31: 391-406.

Chapter 10

Managing the Soil Habitat for Enhanced Biological Fertility

M. Jill Clapperton[1], K. Yin Chan[2], and Frank J. Larney[1]
[1] Agriculture and Agri-Food Canada, Lethbridge Research Centre, P.O. Box 3000, Lethbridge Alberta T1J 4B1 Canada
[2] Wagga Wagga Agricultural Institute, NSW Agriculture, PMB, Wagga Wagga, NSW 2650, Australia.

1.   INTRODUCTION

Soil is a complex inorganic and organic matrix, the habitat for a highly diverse community of microorganisms, fauna and plants, all of which affect the fertility and hence the primary productivity of the ecosystem that they inhabit.   Soil fertility is largely dependent on the processing of organic substrates - soil organic matter (SOM) - through the soil food-web (Swift 1997).   The maintenance of a suitable soil habitat with adequate quality and quantity of organic substrates is therefore critical for microbial communities (Elliot *et al.* 1988, Young and Ritz 2000) and faunal communities (Tian *et al.* 1993, Lavelle *et al.* 1998, Yeates 1999) to cycle nutrients and make them available to plants.   In agriculture we modify the soil habitat and so influence the ability of the soil ecosystem to provide essential services such as decomposition and nutrient cycling for food and fibre production (Constanza *et al.* 1997).   Management practices such as tillage can make it harder or easier for soil organisms to cycle nutrients.   Integrating cropping practices that increase plant diversity (such as inter-, mixed- and cover-

L.K. Abbott & D.V. Murphy (eds) Soil Biological Fertility - A Key to Sustainable Land Use in Agriculture.
203-224
© 2007 *Springer.*

cropping and agro-forestry) and diversified crop rotations (including annual, biennial and perennial crops) together with conservation tillage or no tillage, further increases the potential for managing the soil for enhanced biological fertility by varying the quantity and quality of plant litter (including roots). The chemical and physical quality of the residues affect the populations and diversity of soil biota (Swift *et al.* 1979, Tian *et al.* 1993), the rate of decomposition (Palm and Sanchez 1991, Tian *et al.* 1992) and the subsequent movement of nutrients through the decomposer subsystem (Bardgett *et al.* 1999).

We argue that farmers and other resource managers should consider adopting conservation tillage or no tillage together with diversified cropping practices and crop rotations to manage the soil habitat for enhanced biological fertility. These practices are among the keys to sustaining agriculture as we strive to continue to maintain yields, produce nutritious food, use inputs more effectively and efficiently, conserve natural resources including biodiversity, and reduce the environmental consequences of agriculture.

2.   A SUITABLE HABITAT IS THE KEY TO ENHANCING SOIL BIOLOGICAL FERTILITY

The structural stability of the habitat space and an adequate supply of plant residues and SOM are the foundations for enhancing soil biological fertility. The soil structure can partition resource patches and isolate components of the biological community, altering predator - prey relationships. The soil pore network determines the spatial and temporal distribution of substrates and soil biota, and provides flow paths for solutes and gases. Within the habitable pore space network the spatial distribution of water films and available organic matter have the ultimate control over microbially-mediated soil processes (Young and Ritz 2000). It was suggested that these processes were being controlled at a small-scale by tillage, exploring roots, and habitat modifications by macrofauna (earthworms, termites and ants), demonstrating the importance of the interrelationships between soil biota, plants and soil management practices in terms of regulating soil processes.

The geometry and stability of the soil habitat is mostly defined by the actions of soil biota (Foster 1988, Tisdall and Oades 1982, Lavelle 2000). Ecosystem-engineering organisms modulate soil processes affecting the suitability of the habitat for other organisms including plants (Anderson 1995). Coleman and Crossley (1996) suggested that from an evolutionary and successional standpoint, the properties of an individual that improves the environment or increases the reproductive success of that individual are also likely to benefit other soil organisms. Soil organisms are continuously modifying the habitat to their advantage; earthworms do this at a macro scale and soil microorganisms at a micro scale. For example, the activities of soil invertebrates that lead to increased stability of SOM have evolved to a certain extent from the benefits of increasing the suitability of the soil they inhabit (Wolters 2000). It is likely that more time should be taken to consider the consequences of the habitat modifying behaviour of organisms on plant-organism-soil interactions (Waid 1997), and how this is affected by cropping and soil management practices.

Plants can also affect soil biota and the suitability of the soil habitat by inputs of above- and below-ground residues and root exudates, and by the removal and redistribution of water and mineral nutrients through root uptake. Clearly, any factor or soil management technique that changes the quantity and/or quality of organic material going into the soil, as either residue or root exudates, will effectively change the soil biological community. This will be followed by a myriad of consequences (both negative and positive) for the soil habitat, many of which could limit plant nutrient uptake and growth, and the quantity and quality of plant residues returned to the soil. Thus, management practices that preserve the integrity of the partnership between plants, soil biota and the soil as a habitat will enhance soil biological fertility.

3. ASSESSING A SUITABLE HABITAT AND ENHANCED SOIL BIOLOGICAL FERTILITY

A suitable habitat for any living organism is characterised by its physical structure and biotic properties. The primary indicator of sustainable land management is the assessment of soil health and the direction of the change with time (Karlen *et al.* 1997). There is a need for reliable and easily measured methods of assessing changes in soil structure and biotic properties that reflect soil habitat changes associated with altered cropping and soil management practices. The measurement of various soil biological properties to evaluate soil health has been proposed (Pankhurst *et al.* 1995), and the sensitivity and importance of including microbial and biochemical analyses as soil fertility indicators should not be ignored (Visser and Parkinson 1992, Brookes 1995, Svensson and Pell 2001). The use of soil fauna as indicators of soil quality has also been reviewed (Linden *et al.* 1994, Lobry de Brun 1997). There is probably a suite of biological indicators as opposed to one key organism or measurement that is most likely to reflect soil health. It is also likely that the organisms and measurements included in an indicator suite would differ between soils and climatic zones. In some cases it may be more appropriate to look for organisms that are sensitive to change in their environment, and measure change as opposed to health or quality (Day 1990).

We have chosen to use shifts and changes in earthworm populations and/or species diversity throughout this chapter to demonstrate the suitability of a habitat and the potential for crop benefit and health from enhanced soil biological fertility. However, we acknowledge that excellent arguments can also be made for considering populations and species diversity of protozoa (Foissner 1997, Bamforth 1999, Griffiths *et al*. 2001), nematodes (Porazinska *et al*. 1999, Neher and Barbercheck 1999), and enchytraeids (van Vliet *et al*. 1995) to also demonstrate the effects of soil management practices on soil biological fertility.

4.   EARTHWORMS AS INDICATORS

It has been argued that we need to pay more attention to the effects of tillage and cropping practices on earthworms if we are to build and maintain soil biological nutrient cycling and the soil habitat structural stability required for enhanced biological fertility (Springett *et al*. 1992). Indeed, earthworms are considered to be ecosystem engineers (Lavelle 1997, Lavelle 2000, Anderson 2000), and have been shown to be indicators of soil health and plant growth as well as beneficial land reclamation (Linden *et al*. 1994, Pankhurst *et al*. 1995, Buckerfield *et al*. 1997). They can modify the physical, chemical, and biological properties of soil, and contribute to nutrient cycling (Blair *et al*. 1995), soil aeration and water infiltration (Ketterings *et al*. 1997). Studies have shown that earthworms can also affect the species composition of microorganisms, including protozoa in the soil and around the roots of plants (Gunn and Cheritt 1993, Stephens *et al*. 1994, Doube *et al*. 1994, Bonkowski and Schaffer 1997). Such interactions are important for nutrient cycling and plant productivity (Brown 1995).

The lining of the earthworm burrow (also known as the drilosphere) has been found to have higher populations of nitrifying bacteria than the soil outside the burrow (Parkin and Berry 1999). The increased nitrogen available in the drilosphere could preferentially encourage plant roots to explore earthworm channels. The demonstrated relationship between plant roots and earthworm burrows is complex (Springett *et al*. 1994), with some plant roots preferentially exploring earthworm burrows, while other plant roots determine the distribution of earthworm burrows (Springett and Gray 1997). High earthworm populations are not merely associated with favourable soil fertility but actively build and maintain soil fertility in tropical ecosystems (Hauser *et al*. 1997, Lavelle *et al*. 1998). It is likely that earthworms have much the same effect on soils in other climatic zones. However, there are highly productive soils around the world where earthworms do not exist, possibly because of glaciation, physical barriers to migration or for reasons that are yet unknown. In these soils it is possible that other macro invertebrates such as enchytraeids (van Vliet *et al*. 1995), microarthropods (Behan-Pelletier 1998, Clapperton *et al*. 2002), protozoa and/or nematodes could be valuable indicators of sustainable land management practices. Soil management practices that build populations and diversity of earthworms or other soil fauna that modulate the soil ecosystem are likely to have far-reaching consequences on soil health and productivity.

5. REDUCED SOIL DISTURBANCE TO MANAGE THE INTEGRITY OF THE HABITAT

For thousands of years humans have manipulated the soil in various ways to improve the conditions for crop growth, Tull (1751) advocated modifying the soil physical properties with tillage as a reasonable way to enhance soil fertility and increase yields. However, experiments comparing tillage practices have shown that plant productivity is not related to the tillage implement that was used but rather to the soil environment which it created (Carter 1994). In the last two decades, a Worldwide revolution in tillage practices has taken place. Conservation tillage (ie. minimum tillage or reduced tillage) as defined by Carter (1994) and no tillage are rapidly becoming the norm, and conventional tillage, which relies on intensive soil manipulation (inversion and mixing), has lost favour. Conservation tillage and no tillage were initially adopted for their role in reducing soil degradation by wind and water erosion. In addition, no tillage protects soil from biological degradation (Aslam *et al*. 1999). The benefits of conservation tillage and no tillage to soil biological properties have been well documented (Hendrix *et al*. 1986, Doran and Linn 1994, Beare 1997, Young and Ritz 2000, Ferreira *et al*. 2000). No tillage and to a lesser extent conservation tillage retain the soil surface layers which contain those aggregates richest in SOM, preserving the soil biological component (Dick *et al*. 1997, Peters *et al*. 1997) important to soil fertility and crop productivity.

Conventional tillage affects the placement of residues, collapses the pores and tunnels that were constructed by soil animals and plant roots, and changes the water holding, gas and nutrient exchange capacities of the soil. Conservation tillage (Carter 1994) and particularly no tillage (direct-seeding) create soils that are favourable habitats for soil- and litter-dwelling organisms. Significantly greater earthworm populations have also been reported in soils under no tillage compared with conventional-tillage in Australia (Buckerfield 1992), New Zealand (Francis and Knight 1993), Canada (Clapperton *et al.* 1997), United States of America (Parmelee *et al.* 1990), Finland (Nuutinen 1992), Great Britain (Edwards and Lofty 1982) and Germany (Tebrugge and During 1999) demonstrating the generally positive response of earthworm populations to reduced soil disturbance. Soils with less tillage also have buffered temperatures, improved structure, increased organic matter content, more biologically active and diverse biotic communities, higher nutrient loading capacities, and release nutrients gradually and continuously (Alvarez and Alvarez 2000, Beare *et al.* 1994, Doran and Linn 1994, Angers *et al.* 1993, Arshad *et al.* 1990, Hendrix *et al.* 1986, House *et al.* 1984).

5.1 Making the Transition to a Reduced Tillage System

Conservation tillage and particularly no tillage have been considered the key to enhancing agricultural sustainability (Papendick and Parr 1997), and the benefit to soil health has been documented. However, there are social, economic and agronomic limitations to farmers adopting conservation tillage. In making the change from conventional tillage to conservation tillage or no tillage, farmers must begin applying some of the principles of integrated weed, disease and insect management, and overcome yield-limiting factors that are related to rebuilding the habitat. In humid regions, high yields of crop residues can cause problems for seed germination and establishment. However, this is now considered to be a mechanical constraint (Carter 1994). Decreased plant growth and vigour have also been reported under no tillage systems because of water soluble toxins from the residue and/or toxins released as a consequence of microbial decomposition (Kimber 1967). Alternating the sequence of crops in the rotation can ameliorate these effects from residues (Wolfe and Eckert 1999). This means that reducing the amount of disturbance alone is not sufficient to fully exploit soil biological fertility.

Farmers in the United States of America reported that many of the yield-limiting problems in the first years of the transition to a no tillage system were temporary (Papendick and Parr 1997). In German agriculture, it was suggested that conservation tillage would only be likely to replace ploughing if there were appropriate machinery, diversified crop rotations, and an increased awareness of plant health (Tebrugge and During 1999). There is a documented need for crop rotation in conservation tillage and especially in no tillage systems to provide the soil biological activity to suppress the build-up of rhizoorganisms deleterious to plants, and provide sufficient biodiversity to maintain optimum soil and crop productivity regardless of climate and soil type (Carter 1994). Indeed, researchers increasingly agree that crop rotations and cover crops can be used in conservation tillage and no tillage systems to maintain yields (Papendick and Parr 1997, Hao *et al.* 2000, Tebrugge and During 1999, Drinkwater *et al.* 2000), reduce weed populations (Liebman and Dyck 1993, Blackshaw *et al.* 2000), and increase plant health (Vargas-Ayala *et al.* 2000) to reduce the agronomic risk associated with the transition to no tillage.

6. PLANTS DIRECTLY AND INDIRECTLY INFLUENCE THE SUITABILITY OF THE SOIL HABITAT FOR ENHANCED BIOLOGICAL FERTILITY

Crop rotation presents soil organisms with varied living conditions and a greater variety of substrates. Plants regulate the activities of soil biota (Swift and Anderson 1996) both directly and indirectly. The roots modify the soil structure, and alter the vertical distribution of nutrients, water and soil organisms. The quantity and quality of above- and below- ground residues determines the composition of microbial and faunal communities affecting the formation of soil aggregates and stabilising or destabilising SOM. Populations and the activities of earthworms and other soil 'ecosystem engineers' are strongly influenced by residues, root exudates, and products of decomposition. Therefore, diversified crop rotations are essential for creating a suitable environment for enhanced biological fertility.

6.1 Root Architecture and Root Residues

Diversified crop rotations present a range of root architectures. Root architecture is an important element affecting plant nutrient uptake. The patterns of root response to soil factors such as soil physical structure can vary depending on the plant species and even different genotypes and cultivars within the same species (Zobel 1992). Root ramification and decaying roots add more continuity to the network of soil pores. There is a relationship between the density and distribution of roots and the size and density of aggregates. The length of root in aggregates decreased exponentially with increasing aggregate density, and root growth shifted from within micropores to macropores with increasing aggregate size (de Frietas *et al.* 1999). Root distribution can affect nematode distribution, and root diameter can determine nematode species composition (Yeates 1987). The quantity of carbon (C) allocated to structural biomass, respiration and exudation are also influenced by root architecture (Nielsen *et al*. 1994). This can have an effect on the microbial populations and mineralisation of SOM, as plant structural materials and exudates have very different rates of decomposition.

Roots left in the soil are often ignored source of organic matter, and root architecture and biomass vary dramatically between crop species (Zobel 1975), affecting aggregate stability (Tisdall and Oades 1982), habitat and nutrient dynamics (Jobbagy and Jackson 2001). Heal *et al*. (1997) pointed to the important contribution that roots make to C flow in the soil, and complained that there was little research aimed at determining how root residues contribute to replenishing SOM in arable cropping systems. Recently, it has been reported that root-derived materials are more rapidly occluded by aggregates than shoot-derived residues, and are more likely to contribute to humic materials where roots are concentrated (Wander and Yang 2000). These researchers further concluded that root derived soil organic C in occluded particulate organic matter and humic fractions were more likely to be persistent in the long-term compared with shoot derived soil organic C. There is a demonstrated need to include the contribution of roots to organic matter dynamics and nutrient cycling in agroecosystems.

6.2   Living Roots

Differences in the rates of litter decomposition and nutrient cycling have been reported in the presence of living roots (Bottner *et al*. 1999, Pare *et al*. 2000). Pare *et al*. (2000) showed that 38% of the ^{15}N in alfalfa or lucerne (*Medicago sativa* L.) shoot residues were mineralised when maize plants were present compared with 23% when no plants were included. Interestingly, in the early rapid decomposition stage, competition between plants and microbes for inorganic N reduced the ^{14}C mineralisation of crop residues and decreased plant productivity (Bottner *et al*. 1999). In the same study after 3-6 months, the presence of living roots stimulated ^{14}C mineralisation in the remaining more recalcitrant residues. These results reinforce the importance of roots as a source of SOM, and show that roots possibly have some control over the recycling of nutrients. Therefore, it may be possible to manipulate the plant species and sequence of crops in a rotation, to synchronise the nutrient release from residues with subsequent crop uptake.

6.3   Root Exudates

Root exudates are probably the most labile form of SOM and one of the determining factors in maintaining soil fertility and structural stability in agricultural soils. Root age and type, and the nutritional status of the plant can alter the quality and quantity of root exudates (Yang and Crowley 2000). Nutrient availability in the rhizosphere is in turn affected by the species composition and activities of the biotic community.

Root exudates are the high energy source substrates that support the abundant microbial community in and around the rhizosphere. Microbial activity in the rhizosphere contributes directly and indirectly to plant nutrition by fixing and cycling N, solubilising P (Clarholm 1994), and binding soil particles into larger water stable aggregates (Lee and Foster 1991). Bacteria- and fungal- feeding protozoa and nematodes attracted to the rhizosphere can make significantly more nutrients available to the plant. For example, non-parasitic protozoa and nematodes have been shown to increase N content and shoot biomass (Neher and Barbercheck 1999). The intense biotic activity in the rhizosphere also attracts other larger fauna such as earthworms (Binet *et al.* 1997), the activities of which subsequently modify the soil habitat, and further increase N-mineralisation (Willems *et al.* 1996).

Evidence clearly supports the possibility that plants can regulate both the quantity and quality of C substrate in the rhizosphere as exudates, and affect plant-specific colonisation by rhizosphere microorganisms (Nehl *et al.* 1996). This could then affect plant health because the activities of individual colonies of rhizobacteria can be positive, negative, or neutral to plant growth, depending on habitat characteristics, host genotype, and mycorrhizal status (Nehl *et al.* 1996). The microbial community associated with the rhizosphere of plants colonised by mycorrhizae has been shown to be significantly different from that of non-mycorrhizal plants (Ames *et al.* 1984), this is likely because plants colonised by mycorrhizae partition more photosynthate to the roots (Wang *et al.* 1989, Clapperton and Reid 1992). It has also been shown that mycorrhizal plants can have a higher proportion of amino and organic acids in the roots compared with roots of non-mycorrhizal plants (Clapperton and Reid 1992). A thorough review of the interactions between root exudation, microbial activity and nutrient cycling is provided by Grayston *et al.* (1996).

6.4   Crop Residues

The crop species used, and the sequence of these crops in rotation can affect the quantity and quality of residues. It is well established that the chemical composition and lignin content of plant residues varies with species. This can limit the population and diversity of decomposer organisms, altering the rate of decomposition and soil nutrient cycling (Tian *et al.* 1992, Tian *et al.* 1993, Watkins and Barraclough 1996, Cookson *et al.* 1998). It was suggested that soil invertebrates preferentially ingest high-quality residues (Brussaard 1998). Later, Tian *et al.* (1997) demonstrated that invertebrates significantly affect the turn-over of low quality residues by stimulating microbial activity. Thus, the ability of invertebrates to destabilise and stabilise SOM can be highly dependent on residue quality.

6.5 Summary

In summary, plants provide the substrate, as residues and exudates, for soil organisms to stabilise aggregates and recycle soil nutrients. It is the quantity and quality of crop residues that largely determine the population and diversity of soil biota. Together, plants and soil biota continuously modify the soil as a habitat to further enhance nutrient cycling and plant growth. Therefore, any factor or agricultural practice that changes the amount and/or quality of organic material going into the soil will alter the activities and population dynamics of the soil biota. This in turn can have both short- and long- term positive or negative consequences for plant health and productivity. It is unfortunate that plant residues are often viewed as a nuisance or a medium that harbours disease rather than a resource for soil biota to recycle.

7. CROPPING PRACTICES THAT RETAIN ADEQUATE SOM ARE THE KEY TO REBUILDING THE HABITAT UNDER REDUCED TILLAGE

In order to restore and enhance soil biological fertility in soils that have been conventionally managed, there is a need to reduce the amount of tillage, and supply the optimum amount and quality of residue required to fuel the increased biological activity. Cropping practices that include pastures and perennial crops (including legumes), vary the quantity and quality of SOM and restore populations of soil biota and habitat stability.

For example, a continuous source of fresh plant litter is required to maintain populations and diversity of litter macrofauna (Vohland and Schroth 1999) including earthworms (Lavelle *et al.* 1998). Saprophages consume approximately 15-30% of the annual input of organic matter and Oligochaetes like earthworms and enchytraeids take the biggest share (Wolters 2000). Earthworms stimulate microbial activity and can accelerate the turnover and loss of C if adequate quantities of litter and SOM are not maintained. Therefore, the cost of earthworm activity in terms of organic C needs to be accounted for in agroecosystems (Lavelle *et al.* 1998). This example also illustrates the need to monitor and manage the quantity and quality of SOM going into agroecosystems to maintain biological fertility.

7.1 Pasture and Perennial Crops

In a long-term cropping study, Wardle *et al.* (1999) showed that soil arthropods were most responsive to cropping and soil management practices that affected the nature and quality of the substrate input. Including a perennial crop or pasture phase in the rotation has been shown to restore soil health and the habitat (Paustian *et al.* 1990, Gebhart *et al.* 1994). Short-term pasture (up to 5 yrs) can have a positive effect on the quantity and quality of SOM which is associated with benefits to N fertility and soil structural stability (Haynes 1999).

Earthworm population and species diversity also increase significantly under pasture (Baker *et al.* 1999) and pasture phases in the rotation (Fraser *et al.* 1996, Haynes 1999). The increase in population and diversity in all cases was attributed to the increase in organic matter input under pasture compared with intensive arable cropping. It appears that maintaining an adequate level of SOM can increase the resistance and resilience of soil organisms and processes to disturbance. The pasture or perennial phase in a rotation also represents a cropping phase with reduced soil disturbance that would benefit soil organisms, much like no tillage.

7.2 Cover Crops

The use of cover crops and living mulches in rotation is an effective cropping practice to increase SOM, and depending on the plant species used they can control weeds (Blackshaw *et al.* 2000) and insects too (Vandermeer 1995). However, microbial metabolic diversity has been shown to increase more under pastures and perennial crop phases than under annual cropping sequences including legume cover crops because of tillage (Bending *et al.* 2000). Unfortunately, cover cropping is mostly associated with extensive tillage to incorporate the residues (green manuring). On the contrary, tillage is not always necessary for maximum biological and nutrient cycling benefit (Mohr *et al.* 1998, Drinkwater *et al.* 2000).

Increasing the diversity of residues and quantity of SOM using legume and cereal cover crops under reduced tillage, has the potential to increase the population and diversity of soil biota. Indeed, high densities of microarthropods have been associated with the higher SOM inputs from cover crops, and clover under-sown cereals (Axelsen and Kristensen 2000). Still, there is a paucity of information with respect to interactions between cover crops, soil biota and soil physical and chemical properties.

This lack of information continues to make it difficult to predict where and when cover crops function best (Vandermeer 1995).

7.3   Agroforestry

The added leaf litter and organic substrate from tree roots, combined with crop roots in agroforestry practices, has been shown to increase SOM, stimulating soil microbial activity and increasing soil nutrient pools (Chander *et al.* 1998, Seiter *et al.* 1999). Fine tree roots within alley cropping systems can also significantly influence nutrient cycling because their decomposition releases N and P faster than that of leaves (Jose *et al.* 2000). Tree prunings in tropical ecosystems, unlike temperate ecosystems, can significantly increase SOM content and nutrient cycling (Seiter *et al.* 1999). The maintenance of semi-natural habitats such as strips of trees have the added benefit of harbouring bacteria, fungi (Seiter *et al.* 1999) and beneficial insects (Pfiffner and Luka 2000).

8.   CROP BREEDING FOR BIOLOGICALLY ACTIVE SOILS UNDER CONSERVATION TILLAGE

As more farmers and resource managers consider the transition to conservation tillage and no tillage, they must also contemplate the associated transition to low-input agriculture. Consumers and the public continue to demand food that has been produced in an environmentally acceptable manner with less chemical input. The availability of crop varieties specifically bred to extract nutrients more efficiently and effectively in low input or no chemcial input reduced tillage conditions, would likely be an advantage to producers given the differences in the ways nutrients are recycled between tilled and no tillage systems (Beare 1997). No tillage systems tend to have lower mineralisation and more retained N, and the activities of the soil biota tend to be more seasonally dependent compared with tillage systems (Beare 1997). Ideally, these crops would extract and use mineral nutrients that were made available through soil food webs, and be adapted to inter- and mixed- cropping.

Crop breeding has often compromised root growth for shoot growth and seed production (Zobel 1992, Klepper 1992). Klepper (1992) concluded that crop breeding programs need to consider designing crop rooting systems with traits that would enhance rhizosphere processes. She also suggested that agricultural managers think more about using mixed cropping systems and more diversified rotations to manipulate root system distributions in the soil profile, optimising the capacities of roots to obtain water and nutrients. For example, when we create an above-ground plant canopy structure with inter- and mixed cropping and agro-forestry practices, we also create a root canopy structure (Klepper 1992). Root architecture can shape vertical nutrient profiles, and nutrient distribution patterns along the root (Jobbagy and Jackson 2001). This change in the vertical stratification of roots and nutrients would likely cause a complimentary stratification of SOM affecting the spatial distribution of rhizosphere communities. Thus, the soil-food web becomes more vertically stratified, as do the soil aggregates that provide a more suitable soil habitat for root growth and nutrient uptake. In order to take full advantage of structured root canopies, we need more information linking plant genetics to root architecture and the amount and quality of root exudates. Clearly, this information is critical if we are to manipulate the rhizosphere for crop productivity and also enhance desirable soil properties.

Root exudation has been studied in some modern cereal crops and pasture grasses. However, there are few studies that have compared modern lines of agricultural crops to their ancestors, so we do not know how or if crop selection has changed assimilate partitioning between shoots, roots and the rhizosphere (Hoffman and Carroll 1995). This has implications for the use of transgenic crops (Altieri 2000), of which we know even less. It has been reported that wild type wheat and tetraploid wheat transport proportionally more assimilates to the roots after anthesis than hexaploid wheat (Hoffman and Carroll 1995). The land races from which modern wheat varieties were bred have also been shown to have a higher dependency on symbionts like mycorrhizae compared with more recent plant varieties (Hetrick and Bloom 1983).

Most breeding lines are grown under optimal conditions, where competition between plants for nutrients, water, space and light does not exist. Therefore, symbionts no longer afford these plants a competitive advantage. Parke and Kaeppler (2000) concluded that plant breeders should evaluate the contribution of mycorrhizal fungi to nutrient uptake, drought and disease resistance when selecting germplasm, and that ultimately the genes responsible for mycorrhizal colonisation and responsiveness should be mapped and used when developing new cultivars. This would allow us to exploit both the mycorrhizal symbiosis and the associated benefits to rhizosphere processes, including enhanced plant nutrient uptake and increased habitat structural stability in low input agricultural systems. There needs to be a concerted effort to breed crops with root characteristics and properties that are: adapted to minimum soil disturbance, responsive and encourage beneficial microbial associations, and produce the minimum required amount of root and shoot biomass to maintain adequate levels of SOM.

9.   A WHOLE SYSTEM PERSPECTIVE ON SOIL HABITAT MANAGEMENT

There is ample scientific literature advocating agroecosystem management strategies (organic, biodynamic, low-input, alternative) that foster a more ecological approach to agriculture (see Chapter 12 this volume). These systems all incorporate practices that maintain or increase SOM inputs to enhance biological activity and optimise nutrient cycling. Knowledge of the structure and function of below-ground food webs and their temporal and spatial variation has been considered crucial to understanding the potential for agricultural practices to manipulate and sustain soil fertility and productivity (Beare 1997).

Ideally, agroecosystems should be managed to maintain the structural integrity of the habitat, increase SOM, and optimise the C:N ratios in SOM using cover crops and/or crop sequence to synchronise nutrient release with plant uptake. The quantity and quality of organic matter input and soil disturbance are the factors that most affect soil biota (Swift 1994), and soil biota play a key part in the processes of decomposition and nutrient cycling. This makes an understanding of the relationship between the spatial and temporal abundance and diversity of biotic communities with their effects on habitat, SOM, and nutrient cycling critical for designing soil management practices (Lavelle 2000). The agroecosystem models described by Lavelle (2000) placed importance on soil structure, but focused on biogenic structures, or the voids and organo- mineral structures (e.g. casts and faecal pellets) produced by soil invertebrate engineers, as the components of soil structure that promote a more suitable habitat for plant growth.

It has also been argued that we should consider modelling soil and crop management practices in a way that would allow agricultural soils to more closely resemble soils in natural ecosystems (Soule and Piper 1992, Piper 1999) and optimise the nutrient cycling and soil habitat building activities of soil biota (Neher 1999). This would mean reduced tillage and pesticide use, and more emphasis on perennial and SOM-building crops in the rotation, application of manure and compost for increasing SOM, and synchronising nutrient release and water availability with plant demand (Vandermeer 1995, Neher 1999).

10. CONCLUSION

Agricultural practices that maintain the integrity of the soil habitat and the optimum amount of diverse residues to sustain soil biota will likely enhance soil biological fertility. The availability and immobilisation of nutrients associated with the stability of SOM within an agroecosystem is largely a function of cropping and soil management practices. The ability of farmers and resource managers to successfully make the transition to an agroecosystem relying more on soil biological fertility, will require a greater understanding of rhizosphere processes and how soil food-webs function in these agroecosystems.

11. ACKNOWLEDGEMENTS

We thank Megan Ryan, Henry Janzen, and David Pearce for their useful comments and discussions on earlier versions of this chapter.

12. REFERENCES

Altieri M A 2000 The ecological impacts of transgenic crops on agroecosytem health. Ecosystem Health 6: 13-23.

Alvarez C R and Alvarez R 2000 Short-term effects of tillage systems on active soil microbial biomass. Biology and Fertility of Soils 31: 157-161.

Ames R N, Reid C P P and Ingham E R 1984 Rhizosphere bacterial population responses to root colonisation by vesicular-arbuscular mycorrhizal fungus. New Phytologist 96: 555 - 563.

Anderson J M 1995 Soil organisms as engineers: microsite modulation of macroscale processes. *In:* Linking Species and Ecosystems. C J Jones and J H Lawton (eds.) pp. 94-106. Chapman and Hall. New York, USA.

Anderson J M 2000 Food web functioning and ecosystem processes: problems and perceptions of scaling. *In:* Invertebrates as Webmasters in Ecosystems. D C Coleman and P F Hendrix (eds.) pp. 3-24. CAB International Publishing. Wallingford, UK.

Angers D A, Samson N and Légére A 1993 Early changes in water-stable aggregation induced by rotation and tillage in a soil under barley production. Canadian Journal of Soil Science 73: 51-59.

Arshad M A, Schnitzer M, Angers D A and Ripmeester J A 1990 Effects of till vs no-till on the quality of soil organic matter. Soil Biology and Biochemistry 22: 595-599.

Aslam T, Choudhary M A and Saggar S 1999 Tillage impacts on soil microbial biomass C, N and P, earthworms and agronomy after two years of cropping following permanent pasture in New Zealand. Soil and Tillage Research 51: 103-111.

Axelsen J A and Kristensen K T 2000 Collembola and mites in plots fertilised with different types of green manure. Pedobiologia 44: 556-566.

Baker G H, Carter P J and Barrett V J 1999 Influence of earthworms, *Aporrectodea* spp. (*Lumbricidae*), on pasture production in south-eastern Australia. Australian Journal of Agricultural Research 50: 1247-1257.

Bamforth S S 1999 Soil microfauna: diversity and applications of protozoan in soil. *In:* Biodiversity in Agroecosystems. W W Collins and C O Qualset (eds.) pp. 19-25. CRC Press. Boca Raton, USA, .

Bardgett R D, Mawdsley J L, Edwards S, Hobbs P J, Rodwell J S and Davies W J 1999 Plant species and nitrogen effects on soil biological properties of temperate upland grasslands. Functional Ecology 13: 650-660.

Beare M H, Cabrera M L, Hendrix P E and Coleman D C 1994 Aggregate-protected and unprotected organic matter pools in conventional and no-tillage soils. Soil Science Society of America Journal 58: 787-795.

Beare M H 1997 Fungal and bacterial pathways of organic matter decomposition and nitrogen mineralization in arable soils. *In:* Soil Ecology in Sustainable Agricultural Systems. L Brussaard and R Ferrera-Cerrato (eds.) pp. 37-70. CRC Press. Boca Raton, USA.

Behan-Pelletier V M 1998 Orbatid mite biodiversity in agroecosystems: role for bioindication. Agriculture, Ecosystems and Environment 74: 411-423.

Bending G D, Putland C and Rayns F 2000 Changes in microbial community metabolism and labile organic matter fractions as early indicators of the impact of management on soil biological quality. Biological and Fertility of Soils 31: 78-84.

Blackshaw R E, Moyer J R, Doram R C and Boswell A L 2000 *Meliotus officinalis* cover crop and its residues effectively suppresses weeds during fallow. Weed Science 49: 409-441.

Blair J M, Parmelee R W and Lavelle P 1995 Influences of earthworms on biogeochemistry. *In:* Earthworm Ecology and Biogeography. P F Hendrix (ed.) pp. 127-158. CRC Press Inc. Boca Raton, USA.

Bonkowski M and Schaefer M 1997 Interactions between earthworms and soil protozoa: a trophic component in the soil food web. Soil Biology and Biochemistry 29: 499-502.

Brookes P C 1995 The use of microbial parameters in monitoring soil pollution by heavy metals. Biology and Fertility of Soils 19: 269-279.

Brown G G 1995 How do earthworms affect microfloral and faunal community diversity? Plant and Soil 170: 209-231.

Binet F, Hallaire V and Curmi P 1997 Agricultural practices and the spatial distribution of earthworms in maize fields. Relationships between earthworm abundance, maize plants and soil compaction. Soil Biology and Biochemistry 29: 577-584.

Bottner P, Pansu M and Sallih Z 1999 Modelling the effect of active roots on soil organic matter turnover. Plant and Soil 216: 15-25.

Brussaard L 1998 Soil fauna, guilds, functional groups and ecosystem processes. Applied Soil Ecology 9: 123-135.

Buckerfield J C 1992 Earthworm populations in dryland cropping soils under conservation-tillage in South Australia. Soil Biology and Biochemistry 24: 1667-1672.

Buckerfield J C, Lee K E, Davoren C W and Hannay J N, 1997 Earthworms as indicators of sustainable production in dryland cropping in southern Australia. Soil Biology and Biochemistry 29: 547-554.

Carter M R 1994 Strategies to overcome impediments to adoption of conservation tillage. *In:* Conservation Tillage in Temperate Agroecosystems. M R Carter (ed.) pp. 3-19. CRC Press Inc. USA.

Chander K, Goyal S, Nandal D P and Kapoor K K 1998 Soil organic matter, microbial biomass and enzyme activities in a tropical agroforestry system. Biology and Fertility of Soils 27: 168-172.

Clapperton M J, Kanashiro D A and Behan-Pelletier V M 2002 Changes in abundance and diversity of microarthropods associated with Fescue Prairie grazing regimes. Pedobiologia 46: 496-511.

Clapperton M J, Miller J J, Larney F J, Lindwall C W 1997 Earthworm populations as affected by long - term tillage practices in southern Alberta, Canada. Soil Biology and Biochemistry 29: 631-633.

Clapperton M J and Reid D M 1992 Effects of low-concentration sulphur dioxide fumigation and VA mycorrhizal fungi on ^{14}C-partitioning in *Phleum pratense*. New Phytologist 120: 381-387.

Clarholm M 1994 The microbial loop in soil. *In:* Beyond Biomass. K Ritz, J Dighton and K E Giller (eds.) pp. 221-238. Wiley. USA.

Coleman D C and Crossley Jr D A 1996 Fundamentals of Soil Ecology. pp. 128-139. Academic Press Inc. USA.

Constanza R, d'Arge R, de Groot R, Farber S, Grasso M, Hannon B, Limburg K, Naeem S, O'Neill R V, Paruelo J, Raskin R G, Sutton P and van den Belt M 1997 The value of the World's ecosystem services and natural capitol. Nature 387: 253-259.

Cookson W R, Beare M H and Wilson P E 1998 Effects of prior crop residue management on microbial properties and crop residue decomposition. Applied Soil Ecology 7: 179-188.

Day K 1990 Rapporteur's report of work group: indicators at the species and biochemical level. Environmental Monitoring Assessment 15: 277-290.

de Frietas P L, Zobel R W and Snyder V A 1999 Corn root growth in soil columns with artificially constructed aggregates. Crop Science 39: 725-730.

Dick W A, Edwards W M and McCoy E L 1997 Continuous application of no tillage to Ohio soils: changes in crop yields and organic matter-related soil properties. *In:* Soil Organic Matter in Temperate Agroecosystems. E A Paul, E T Elliot, K Paustian and C V Cole (eds.) pp. 171-182. CRC Press. Boca Raton, USA.

Doran J W and Linn D M 1994 Microbial ecology of conservation management systems. *In:* Soil Biology: Effects on Soil Quality. J L Hatfield and B A Stewart (eds.) pp. 1-27. CRC Press Inc. Boca Raton, USA.

Doube B M, Stephens P M, Davoren C W and Ryder M H 1994 Interactions, beneficial microorganisms, and root pathogens. Applied Soil Ecology 1: 3-10.

Drinkwater L E, Janke R R and Rossoni-Longnecker L 2000 Effects of tillage intensity on nitrogen dynamics and productivity in legume-based grain systems. Plant and Soil 227: 99-113.

Edwards C A and Lofty J R 1982 The effect of direct drilling and minimal cultivation on earthworm populations. Journal of Applied Ecology 19: 723-734.

Elliot E T, Hunt H W and Walter D E 1988 Detrital foodweb interactions in North American grassland ecosystems. Agriculture, Ecosystems and Environment 24: 41-56.

Ferreira M C, Andrade D de S, Chueire L M de O, Takemura S M and Hungria M 2000 Tillage method and crop rotation effects on the populations sizes and diversity of bradyrhizobia nodulating soybean. Soil Biology and Biochemistry 32: 627-637.

Foissner W 1997 Protozoa as bioindicators in agroecosystems, with emphasis on farming practices, biocides, and biodiversity. Agriculture Ecosystems and Environment 62: 93-103.

Foster R C 1988 Microenvironments of soil microorganisms. Biology and Fertility of Soils 6: 189-203.

Francis G S and Knight T L 1993 Long-term effects of conventional and zero-tillage on selected soil properties and crop yields in Canterbury, New Zealand. Soil Tillage Research 26: 193-210.

Fraser P M, Williams P H and Haynes R J 1996 Earthworm species, population size and biomass under different cropping systems across the Canterbury Plains, New Zealand. Applied Soil Ecology 3: 49-57.

Gebhart D L, Johnson H B, Mayeaux H S and Polley H W 1994 The CRP increases soil organic matter. Journal of Soil and Water Conservation 49: 488-492.

Grayston S J, Vaughan D and Jones D 1996 Rhizosphere carbon flow in trees, in comparison with annual plants; the importance of root exudation and its impacts on microbial activity and nutrient availability. Applied Soil Ecology 5: 29-56.

Griffiths B S, Bonkowski M, Roy J and Ritz K 2001. Functional stability, substrate utilisation and biological indicators of soils following environmental impacts. Applied Soil Ecology 16: 49-61.

Gunn A and Cherritt J M 1993 The exploitation of food resources by soil meso and macro invertebrates. Pedobiologia 37: 303-320.

Hao X, Chang C, Larney F J, Nitschelm J and Regitnig P 2000 Effect of minimum tillage and crop sequence on physical properties of irrigated soil in southern Alberta. Soil and Tillage Research 57: 53-60.

Hauser S, Vanlauwe B, Asawalam O D and Norgrove L 1997 Role of earthworms in traditional and improved low-input agricultural systems in West Africa. *I:n* Soil Ecology in Sustainable Agricultural Systems. L Brussaard and R Ferrara-Cerrato (eds.) pp. 113-136. CRC Press Inc. Boca Raton, USA.

Haynes R J 1999 Labile organic matter fractions and aggregate stability under short-term, grass-based leys. Soil Biology and Biochemistry 31: 1821-1830.

Heal O W, Anderson J M and Swift M J 1997 Plant litter quality and decomposition: an overview. *In:* Driven by Nature: Plant Litter Quality and Decomposition. G Cadish and K E Giller (eds.) pp. 67-74. CAB International. UK.

Hendrix P F, Parmelee R W, Crossley D A Jr, Coleman D C, Odum E P and Groffman P 1986 Detritus food webs in conventional and no-tillage agroecosystems. Bioscience 36: 374-380.

Hetrick B A D and Bloom J 1983 Vesicular-arbuscular mycorrhizal fungi associated with native tall grass prairie and cultivated winter wheat. Canadian Journal of Botany 61: 2140-2146.

Hoffman C A and Carroll C R 1995 Can we sustain the biological basis of agriculture? Annual Review of Ecology and Systematics 26: 69-92.

House G J, Stinner B R, Crossley D A Jr, Odum E P and Longdale G W 1984 Nitrogen cycling in conventional and no tillage agroecosystems in the southern Piedmont. Journal of Soil and Water Conservation 39: 194-200.

Jobbagy E G and Jackson R B 2001 The distribution of soil nutrients with depth: global patterns and the imprint of plants. Biogeochemistry 53: 51-77.

Karlen D L, Mausbach M J, Doran J W, Cline R G, Harris R F and Schuman G E 1997 Soil quality: a concept, definition, and framework for evaluation. Soil Science Society of America Journal 61: 4-10.

Ketterings Q M, Blair J M and Marinissen J C Y 1997 Effects of earthworms on soil aggregate stability and carbon and nitrogen storage in a legume cover crop agroecosystem. Soil Biology and Biochemistry 29: 401-408.

Kimber R W L 1967 Phytotoxicity from plant residues. I. The influence of rotted straw on seedling growth. Australian Journal of Agricultural Research 18: 361-374.

Klepper B 1992 Development and growth of crop root systems. Advances in Soil Science 19: 1-26.

Lavelle P 1997 Faunal activities and soil processes: adaptive strategies that determine ecosystem function. Advances in Ecological Research 27: 93-132.

Lavelle P 2000 Ecological challenges for soil science. Soil Science 165: 73-86.

Lavelle P, Pashanasi B, Charpentier F, Gilot C, Rossi J P, Derouard L, Andre J, Ponge J F and Bernier N 1998 Large-scale effects of earthworms on soil organic matter and nutrient dynamics. *In:* Earthworm Ecology. C A Edwards (ed.) pp. 103-122. CRC Press Inc. USA.

Lee K E and Foster R C 1991 Soil fauna and soil structure. Australian Journal of Soil Research 29: 745-775.

Liebman M and Dyck E 1993 Crop rotation and intercropping strategies for weed management. Ecological Applications 3: 92-122.

Linden D R, Hendrix P F, Coleman D C and van Vilet P C J 1994 Faunal indicators of soil quality. *In:* Defining Soil Quality for a Sustainable Environment, Soil Science Society of America Publication 35. J W Doran, D C Coleman, D F Bezdicek and B A Stewart (eds.) pp. 91-106. Soil Science Society of America Inc. USA.

Lobry de Brun L A 1997 The status of soil macrofauna as indicators of soil health to monitor the sustainability of Australian agricultural soils. Ecological Economics 23: 167-178.

Mohr R M, Janzen H H, Bremer E and Entz M H 1998 Fate of symbiotically-fixed $^{15}N_2$ as influenced by method of alfalfa termination. Soil Biology and Biochemistry 30: 1359-1367.

Neher D A 1999 Soil community composition and ecosystem processes - comparing agricultural ecosystems with natural ecosystems. Agroforestry Systems 45: 159-185.

Neher D A and Barbercheck M E 1999 Diversity and function of soil mesofauna. *In:* Biodiversity in Agroecosystems. W W Collins and C O Qualset (eds.) pp. 27-47. CRC Press. Boca Raton, USA, .

Nehl D B, Allen S J and Brown J F 1996 Deleterious rhizobacteria: an integrating perspective. Applied Soil Ecology 5: 1-20.

Nielsen K L, Lynch J P, Jablokow A G and Curtis P S 1994 Carbon cost of root systems: an architectural approach. Plant and Soil 165: 161-169.

Nuutinen V 1992 Earthworm community response to tillage and residue management on different soil types in southern Finland. Soil Tillage Research 23: 221-239.

Palm C A and Sanchez P A 1991 Nitrogen release from leaves of some tropical legumes as affected by their lignin and polyphenolic contents. Soil Biology and Biochemistry 23: 83-88.

Pankhurst C E, Hawke B G, McDonald H J, Kirkby C A, Buckerfield J C, Michelsen P, O'Brien K A, Gupta V V S R and Doube B M 1995 Evaluation of soil biological properties as potential bioindicators of soil health. Australian Journal of Experimental Agriculture 35: 1015-1028.

Papendick R I and Parr J F 1997 No-till farming: The way of the future for a sustainable dryland agriculture. Annals of Arid Zone 36:L 193-208.

Pare T, Gregorich E G and Nelson S D 2000 Mineralisation of nitrogen from crop residues and N recovery by maize inoculated with vesicular-arbuscular mycorrhizal fungi. Plant and Soil 218: 11-20.

Parke J L and Kaeppler S W 2000 Effects of genetic differences among crop species and cultivars upon the arbuscular mycorrhizal symbiosis. *In:* Arbuscular Mycorrhizas: Physiology and Function. Y Kapulnik and D D Douds Jr (eds.) pp. 131-146. Kluwer Academic Publishers. The Netherlands.

Parkin T and Berry E C 1999 Microbial nitrogen transformations in earthworm burrows. Soil Biology and Biochemistry 31: 1765-1771.

Parmelee RW, Beare M H, Cheng W, Hendrix P F, Rider S J, Crossley Jr D A and Coleman D C 1990 Earthworms and enchytraeids in conventional and zero-tillage agroecosystems: a biocide approach to assess their role in organic matter breakdown. Biology and Fertility of Soils 10: 1-10.

Paustian K, Andrén O, Clarholm M, Hansson A C, Johansson G, Largerlöf J, Lindberg T, Pettersson R and Sohlenius B 1990 Carbon budgets of four agroecosystems with annual and perennial crops, with and without N fertilization. Journal of Applied Ecology 27: 60-84.

Peters M M, Wander L S, Saporito L S, Harris G H and Friedman D B 1997 Management impacts on SOM and related soil properties in a long-term farming systems trial in Pennsylvania: 1981-1991. *In:* Soil Organic Matter in Temperate Agroecosystems. E A Paul, E T Elliot, K Paustian and C V Cole (eds.) pp. 183-196. CRC Press Inc. USA.

Pfiffner L and Luka H 2000 Overwintering of arthropods in soils of arable fields and adjacent semi-natural habitats. Agriculture, Ecosystem and Environment 78: 215-222.

Piper J K 1999 Natural systems agriculture. *In:* Biodiversity in Agroecosystems. W W Collins and C O Qualset (eds.) pp. 167-196. CRC Press. Boca Raton, USA.

Porazinska D L, Duncan L W, McSorley R and Graham J H 1999 Nematode communities as indicators of status and processes of a soil ecosystem influenced by agricultural management practices. Applied Soil Ecology 13: 69-86.

Seiter S, Ingham E R and William R D 1999 Dynamics of soil fungal and bacterial biomass in a temperate climate alley cropping system. Applied Soil Ecology 12: 139-147.

Soule J D and Piper J K 1992 Farming in Nature's Image: An Ecological Approach to Agriculture. Island Press. Washington DC, USA.

Springett J A, Gray R A J and Reid J B 1992 Effect of introducing earthworms into horticultural land previously denuded of earthworms. Soil Biology and Biochemistry 24: 1615-1622.

Springett J A, Gray R A J, Reid J B and Petrie R 1994 Deterioration in soil biological and physical properties associated with kiwifruit. Applied Soil Ecology 1: 231-241.

Springett J A and Gray R A J 1997 The interaction between plant roots and earthworm burrows in pasture. Soil Biology and Biochemistry 29: 621-625.

Stephens P M, Davoren C W, Doube B M and Ryder M H 1994 Ability of the Lumbricid earthworms *Aporrectodea rosea* and *Aporrectodea trapezoides* to reduce the severity of take-all under greenhouse and field conditions. Soil Biology and Biochemistry 26: 1291-1297.

Svensson K and Pell M 2001 Soil microbial tests for discriminating beetween different cropping systems and fertiliser regimes. Biology and Fertility of Soils 33: 91-99.

Swift M H 1994 Maintaining the biological status of soil: a key to sustainable land management. *In:* Soil Resilience and Sustainable Land Use. D J Greenland and I Szabolcs (eds.) pp. 33-39. CAB International. Wallingford, UK.

Swift M H 1997 Biological management of soil fertility as a component of sustainable agriculture: perspectives and prospects with particular reference to tropical regions. *In:* Soil Ecology in Sustainable Agricultural Systems. L Brussaard and R Ferrera-Cerrato (eds.) pp. 137-159. CRC Press. Boca Raton, USA.

Swift M J and Anderson J M 1996 Biodiversity and ecosystem function in agricultural systems. *In:* Biodiversity and Ecosystem Function. E D Schulze and H A Mooney (eds.) pp. 15-41. Springer Verlag. Berlin.

Swift M J, Heal O W and Anderson J M 1979 Decomposition in Terrestrial Ecosystems. Blackwell Scientific. UK.

Tebrugge F and During R A 1999 Reducing tillage intensity – a review of results from a long-term study in Germany. Soil Tillage Research 53: 15-28.

Tian G, Kang B T and Brussaard L 1992 Biological effects of plant residues with contrasting chemical compositions under humid tropical conditions: decomposition and nutrient release. Soil Biology and Biochemistry 24: 1051-1060.

Tian G, Brussaard L and Kang B T 1993 Biological effects of plant residues with contrasting chemical compositions under humid tropical conditions: effects on soil fauna. Soil Biology and Biochemistry 25: 731-737.

Tian G, Brussaard L, Kang B T and Swift M J 1997 Soil fauna mediated decomposition of plant residues under constrained environmental and residue quality conditions. *In:* Driven by Nature: Plant Litter Quality and Decomposition. G Cadish and K E Giller (eds.) pp. 125-144. CAB International. Wallingford, UK.

Tisdall J M and Oades J M 1982 Organic matter and water-stable aggregates in soil. Journal of Soil Science 33: 141-163.

Tull J 1751 Horse-hoeing husbandry. 3rd Edition. Miller. London, UK.

Vandermeer J 1995 The ecological basis of alternative agriculture. Annual Review of Ecology and Systematics 26: 201-224.

van Vliet P C J, Beare M H and Coleman D C 1995 Population dynamics and functional roles of Enchytraeidae (Oligochaeta) in hardwood forest and agricultural ecosystems. Plant and Soil 170: 199-207.

Vargas-Ayala R, Rodriguez-Kabana R, Morgan-Jones G, McInroy J A and Kloepper J W 2000 Shifts in soil microflora induced by velvet bean (*Mucuna deeringiana*) in cropping systems to control root-know nematodes. Biological Control 17: 11-22.

Visser S and Parkinson D 1992 Soil biological criteria as indicators of soil quality: soil microorganisms. American Journal of Alternative Agriculture 7: 33-37.

Vohland K and Schroth G 1999 Distribution patterns of the litter macrofauna in agroforestry and monoculture plantations in central Amazonia as affected by plant species management. Applied Soil Ecology 13: 57-68.

Waid J S 1997 Metabiotic interactions in plant litter systems. *In:* Driven by Nature: Plant Litter Quality and Decomposition. G Cadish and K E Giller (eds) pp. 145-156. CAB International. Wallingford, UK.

Wander M M and Yang X 2000. Influence of tillage on the dynamics of loose- and occluded- particulate and humified organic matter fractions. Soil Biology and Biochemistry 32: 1151-1160.

Wang G M, Coleman D C, Freckman D W, Dyer M I, McNaughton S J, Acra M A and Goeschl J D 1989 Carbon partitioning patterns of mycorrhizal versus non-mycorrhizal plants: real time dynamic measurements using $^{11}CO_2$. New Phytologist 112: 489-493.

Wardle D A, Nicholson K S, Bonner K I and Yeates G W 1999 Effects of agricultural intensification on soil-associated arthropod population dynamics, community structure, diversity, and temporal variability over a seven-year period. Soil Biology and Biochemistry 31: 1691-1706.

Watkins N and Barraclough D 1996 Gross rates of N mineralisation associated with the decomposition of plant residues. Soil Biology and Biochemistry 28: 169-175.

Willems J J G M Marinissen J C Y, and Blair J 1996 Effects of earthworms on nitrogen mineralization. Biology and Fertility of Soils 23: 57-63.

Wolfe A M and Eckert D J 1999 Crop sequence and surface residue effects on the performance of no-till corn grown on a poorly drained soil. Agronomy Journal 91: 363-367.

Wolters V 2000 Invertebrate control of soil organic matter stability. Biology and Fertility of Soils 31: 1-19.

Yang C H and Crowley D E 2000. Rhizosphere microbial community structure in relation to root location and plant iron nutritional status. Applied and Environmental Microbiology 66: 345-351.

Yeates G W 1987 How plants affect nematodes. Advances in Ecological Research 17: 63-113.

Yeates G W 1999 Effects of plants on nematode community structure. Annual Review of Plant Pathology 37: 127-149.

Young I M and Ritz K 2000 Tillage, habitat space and function of soil microbes. Soil and Tillage Research 53: 201-213.

Zobel R W 1992 Soil environment constraints on root growth. Advances in Soil Science 19: 27-51.

Zobel R W 1975 The genetics of root development. *In:* The Development and Function of Roots. J G Torrey and D F Clarkson (eds) pp. 261-275. Academic Press. U.K.

Chapter 11

Sustainable Farming Systems and their Impact on Soil Biological Fertility - Some Case Studies

Elizabeth A. Stockdale[1] and W. Richard Cookson[2]

[1] Agriculture and Environment Division, Rothamsted Research, Harpenden, Hertfordshire, AL5 2JQ, United Kingdom.
[2] Centre for Land Rehabilitation, Faculty of Natural and Agricultural Sciences, The University of Western Australia, Crawley, WA 6009, Australia.

1. INTRODUCTION

The term 'sustainable agriculture' is used widely and has embraced a diverse range of issues and objectives, including animal welfare, greater protection of the environment, and the need for farming to support other sectors of the economy such as tourism. Where the principles of sustainable development are applied to agriculture, then farming systems are judged to make a major input to a sustainable economy and society when they concurrently meet the following objectives:

- Produce safe food and non-food products in response to market demands.
- Enable viable livelihoods to be made from land management.
- Operate within biophysical constraints and enable a diverse wildlife.
- Provide environmental and other benefits to the public such as recreation and access.
- Achieve the highest standards of animal health and welfare.

L.K. Abbott & D.V. Murphy (eds) Soil Biological Fertility - A Key to Sustainable Land Use in Agriculture. 225-239.
© 2007 Springer.

- Support the vitality of rural economies and the diversity of rural culture.
- Sustain the resource available for growing food and supplying other public benefits over time.

Every farming system is an agglomeration of structures and practices placed within local environmental constraints (climate and soil). Farming systems are inherently adapted to their location. Farmers combine management of soils with the management of a combination of crops, livestock, wildlife habitats, labour, marketing, storage and processing of products. Recent emphasis on the development of sustainable farming systems has focussed on the application of skills and knowledge in managing biological cycles and their interactions in the farming system within the local institutional and environmental framework (NRC 1993). It is not possible to define a single blueprint approach to sustainable agriculture. No single set of structures and practices can simultaneously achieve all the ecological and socio-economic goals and fit the diverse social and environmental conditions in any region, let alone at a global scale.

In this chapter we first describe some farming systems commonly described as 'sustainable' and then examine a series of case studies that have quantified the impact of these farming systems on soil biological fertility.

2. CHARACTERISTICS OF SUSTAINABLE FARMING SYSTEMS

In practice, the principles of sustainable farming systems are often outlined at a high level, globally or regionally, with the practical implementation at farm-scale left to the farmer and farm advisors. The resulting diversity of agricultural systems, which may describe themselves as 'sustainable', means that it is difficult to identify clear types of farming systems, or indeed to evaluate their sustainability in practice. However, several types of sustainable farming systems have been described (Table 1) and variants of these system types can be recognised in farming systems used by both smallholders and corporations. Sustainable agricultural systems may be distinguished from conventional approaches to agriculture from *inter alia* their focus on system rather than single crop management, increased links between crop and livestock enterprises, increased recognition of the importance of biological processes and the increased cycling of nutrients within the farming system (Figure 1).

Table 1 Key characteristics of several types of sustainable farming systems

LOW EXTERNAL INPUT/ INTEGRATED	ORGANIC	BIODYNAMIC	AGROFORESTRY
Integrating beneficial natural processes into farm practice and seeking to conserve/enhance environmental value.	Creating farming systems which seek to integrate soil, environmental and human health benefits.	Managing farming systems as organisms which optimise soil quality and plant, animal and human health.	Integrating woody and herbaceous perennials into farming systems.
Crop rotation.	Crop rotation and also ideally spatial diversity.	Crop rotation and also ideally spatial diversity.	Spatial diversity of crop types.
Minimum impact tillage.			Variety of cropping and pastoral systems integrated.
Targeted use of fertilisers.	Achievement of self-sufficiency in N through biological N fixation.	Achievement of self-sufficiency in N through biological N fixation.	
Targeted use of pesticides alongside cultural methods for pest, disease and weed control.	Prohibition of synthetic plant treatments and nutrient sources.	Prohibition of synthetic plant treatments and nutrient sources.	
		Use of specific preparations to enhance soil quality and plant life.	
	Extensive livestock management.	Extensive livestock management.	
General principles outlined. Detailed application site and crop specific.	Strong unifying principles. Legislative base throughout much of the world. Detailed application site and crop specific.	Strong unifying principles. Detailed application site and crop specific.	General principles outlined. Detailed application site and crop specific.
Reijntjes *et al.* 1992, El Titi *et al.* 1993, Holland *et al.* 1994	IFOAM 1998, Stockdale *et al.* 2001	Steiner 1924, Koepf *et al.* 1976	Nair 1989, Sanchez *et al.* 1997

a) Organically managed

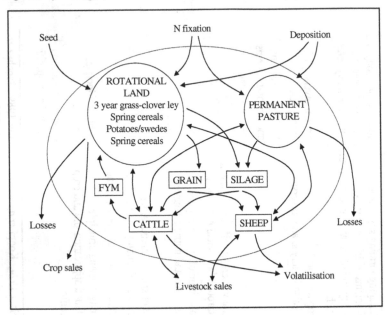

b) Conventionally managed

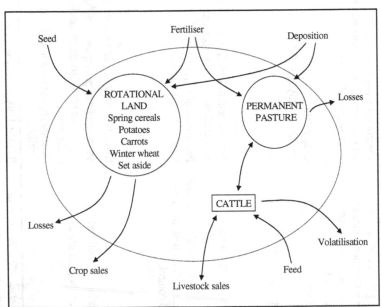

Figure 1 The flows of nitrogen in two contrasting mixed farming systems in Scotland, derived from information presented in Topp *et al.* 2000, showing the increased cycling of nutrients within the farming system in an organic farming system.

2.1 Impact of sustainable farming systems

Low external input and integrated farming systems are widely reported to increase sequestration of carbon (C) in soil and to favour the development of populations of earthworms and other soil fauna (e.g. Bloem *et al.* 1994). Many studies have also shown increases in soil organic matter levels and soil microbial biomass populations in organic and biodynamic farming systems (Stockdale *et al.* 2001). Fewer data are available which describe the impact of agroforestry systems on soil biological fertility. In the following section we present some case studies which examine the impact of these farming systems where long-term trials have included some measurements of soil biological fertility.

3. CASE STUDIES

3.1 Integrated Farming Systems – The Netherlands

The integrated and conventional arable farming systems which were compared in a controlled plot trial had a rotation of winter wheat, sugar beet, spring barley and potatoes (Bloem *et al.* 1994). The integrated system used reduced soil tillage, did not use soil fumigation and reduced inputs of crop protection chemicals and N fertiliser, approximately 40% less, but used green manure crops and received inputs of farmyard manure and mushroom compost during the rotation.

N mineralisation was on average 30% higher in the integrated farming system than in the conventional system; bacterial biomass size and activity also tended to be higher. These differences were attributed to greater inputs of organic matter on the integrated plots (5650 kg organic matter ha^{-1} year^{-1}) compared with the conventional (3200 kg organic matter ha^{-1} year^{-1}); the integrated plots also had 30% higher soil organic matter contents (Bloem *et al.* 1994). However, increased bacterial growth in early spring resulted in increased immobilisation in the integrated system and N mineralisation was delayed by 3-6 weeks compared with the conventional system. Detailed measurements also found a higher biomass of bacterivores (amoebae and nematodes) in the integrated system which may indicate a higher turnover of bacterial biomass (Table 2). This was supported by a higher frequency of divided cells, active bacterial biomass and potential O_2 consumption in the integrated system (Table 2).

Table 2 Soil biological measurements (0-25 cm) from integrated and conventionally managed soil (adapted from Bloem *et al.* 1994).

Measurement	Integrated system	Conventional system
Potential N mineralisation (kg N ha^{-1} week^{-1} cm^{-1})	0.23	0.17
In situ N mineralisation (kg N ha^{-1} week^{-1} cm^{-1})	0.08	0.06
Microbial biomass-N (kg N ha^{-1} cm^{-1})	7.4	4.5
Microbial biomass-C (kg C ha^{-1} cm^{-1})	136	127
Amoebae (kg C ha^{-1} cm^{-1})	0.76	0.46
Bacterivorous nematodes (kg C ha^{-1} cm^{-1})	0.015	0.012
Frequency of dividing-divided cell (%)	3.0	2.4
Potential O$_2$ consumption (kg O$_2$ ha^{-1} week^{-1} cm^{-1})	11	9

3.2 Sustainable Agricultural Systems (SAFS) - America

The Sustainable Agricultural Systems (SAFS) project at the University of California, Davis, is a long-term multidisciplinary study comparing four farm management systems including processing tomatoes (Temple *et al.* 1994). The climate is Mediterranean and irrigation is used in all the agricultural systems. Two and four year conventional rotations, with regional average applications of fertilisers and pesticides, are compared with i) a low-input rotation which includes an over winter cover crop (*Vicia* sp.) and reduced inputs and, ii) an organic rotation which uses winter cover crops, poultry manure and small amounts of seaweed and fish powder as amendments during the rotations. Detailed measurements of the impact of each farming system on soil microbial populations and their activity showed that the total microbial biomass C was higher in organic and low-input systems than either of the conventional rotations (Gunapala and Scow 1998; Table 3), but that the metabolic quotient (respiration per unit biomass) was similar throughout the season in the organic and conventional 4 year rotation. However, the microbial activity per unit biomass peaked in the low input and organic systems after incorporation of the residues of the vetch cover crop, and for a four week period, microbial activity was significantly higher than that on the conventional rotation. Although the management of the systems varied in many ways, the main factors causing differences in the microbial biomass were the size and quality of the inputs of organic C. Carbon inputs in manures had relatively little impact on the size of the microbial population compared with the use of cover crops.

Table 3 Microbial biomass C and arthropod diversity (*) in farming systems within the Sustainable Agricultural Systems (SAFS) project at the University of California, Davis.

Measurement	Management system				Reference
	4 yrs	2 yrs	Low-input	Organic	
Microbial biomass C ($\mu g\ g^{-1}$ soil)	56	47	106	93	Gunapala *et al.* 1998
*Species richness: herbivores	16	n.a.	n.a.	22	
*Species richness: parasitoids	13	n.a.	n.a.	25	Letourneau & Goldstein 2001
*Species richness: predators	9	n.a.	n.a.	13	
*Abundance: parasitoids	119	n.a.	n.a.	223	

Gunapala *et al.* (1998) showed that the differences in the microbial populations measured in the field did not persist under laboratory conditions; microbial communities in the conventional soil were sufficient and active enough to respond similarly when inputs of vetch were made under controlled conditions. However, when soils taken from the field at different times of the season were compared, cover crop decomposition rates were more consistent in soils taken from the organic system. This may suggest a greater abundance and diversity of the microbial community in organically managed soils (Lundquist *et al.* 1999). Bossio *et al.* (1998) also measured differences in microbial diversity (using phospholipid fatty acid profiles) between the conventional and organic plots in the SAFS experiment.

Letourneau and Goldstein (2001) have also shown that the abundance and structure of arthropod communities is important in limiting arthropod damage in similar tomato crop systems (Table 3). Using canonical discriminant analysis, Letourneau and Goldstein (2001) found that although herbivore abundance did not differ, natural enemy abundance and species richness of all functional groups of arthropods (herbivores, predators, parasitoids), were greater in soils from organically managed fields. However, fallow management, surrounding habitat and transplant date of the crop field, not insecticide intensity, explained the major variability in abundance patterns of prominent pests and natural enemies between systems. Thus, there may be close links between the factors which lead to disease suppressive soils (Chapter 8) and the management of other pests.

3.3   Long Term Ecological Research Project - America

In the Long Term Ecological Research (LTER) project farming systems are compared which differ in tillage, source of nitrogen (N), amount and types of chemical inputs used for crop protection, and weed control (Menalled *et al.* 2001). In 1993 and 1999, the influence of these systems on above ground and seedbank weed communities was measured (Menalled *et al.* 2001; Table 4).

Table 4 Weed biomass, weed species density and weed species diversity in the Long Term Ecological Research (LTER) farming systems projects (adapted from Menalled *et al.* (2001)).

Management System	Weed biomass* ($g\ m^{-2}$)	Species density* (species m^{-2})	Shannon diversity index*
Conventional	18a	2.4a	0.18a
No-till	58b	2.9a	0.18a
Low-input	69c	4.7b	0.28b
Organic	109d	6.2c	0.32c

*Differences in letter denotes significant differences between means

Total weed biomass, species density, and diversity were lowest in conventional and no-till systems as herbicide application reduced the density of herbicide susceptible weed species (Table 4). However, while the number of weed seeds significantly increased in the seedbanks of conventional and no-till systems, the seedbank size significantly decreased in the low-input and organic systems (Menalled *et al.* 2001). The weed communities of wheat fields were also distinctly different from those under corn and soybean in low-input and organic systems; no effect of crop type was found in conventional and no-till systems. Menalled *et al.* (2001) concluded that the interaction of tillage, N management and consequent changes in soil properties including soil structure, and the activity of microorganisms and soil micro- and macro-fauna had an impact on the development of the weed seedbank. These results link closely with the concept of developing 'weed suppressive soils' for sustainable farming systems, where management strategies are structured to foster microbial populations in soils that will increase the decomposition of weed seeds and suppress weed growth (Kennedy and Kremer 1996; Quimby *et al.* 2002).

3.4   DOK Trial - Switzerland

A long-term replicated field trial was established in 1978 to compare the effects of biodynamic, organic and conventional farming systems (DOK) at Therwil in Switzerland (Fließbach and Mäder 1997). The conventional system used crop protection chemicals as appropriate and mineral N, P and K fertilisers, both with and without manure inputs. In the conventional system with manure, stacked farmyard manure and slurry are used. In the organic system, farmyard manure is rotted down and the slurry is aerated, whereas in the biodynamic system both farmyard manure and slurry are composted. The plots compared in this case study received manures at a rate equivalent to 1.4 livestock units per hectare.

Fließbach *et al.* (2000) showed that by 1997, soil organic C and microbial biomass C were significantly greater in biodynamic plots than unfertilised plots (Figure 2). The functional diversity of the microbial community, measured in spring, in the biodynamic system was also significantly higher than that of the organic system; both were very significantly greater than in the conventional systems (Fließbach and Mäder 1997). These findings contrast with results from the USA where Carpenter-Boggs *et al.* (2000-a) found that organically and biodynamically managed soils had similar microbial populations; use of biodynamic preparations in soil management did not further increase soil microbial biomass or its activity. However, use of biodynamic preparations while composting dairy manure did lead to different microbial community profiles in the final composts, determined by phospholipid fatty acid profiles (Carpenter-Boggs *et al.* 2000-b). In the DOK trial, the different management of the manures may therefore be the main factor leading to increased microbial activity and diversity in the biodynamic plots. Within the plots of the DOK trial Fließbach *et al.* (2000) also showed that a greater fraction of the straw applied to soil taken from the biodynamic system was mineralised compared with the conventional system, with more straw derived C incorporated into the microbial biomass. Microbial populations with greater diversity required less energy per unit biomass for microbial maintenance. This may result in an increased turnover of organic matter, but with greater conservation of C within the soil system, rather than released as CO_2 (Mäder *et al.* 2002). Enzyme activities (protease, dehydrogenase) were also markedly higher in the biodynamic and organic systems: acid phosphatase activity was also higher within biodynamic (and organic) systems than conventional systems, resulting in greater mineralisation of organic P (Oberson *et al.* 1996).

a) Microbial biomass carbon

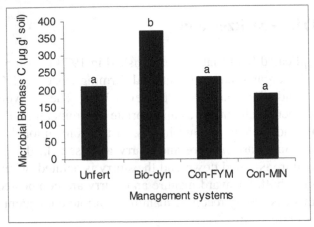

b) Organic carbon

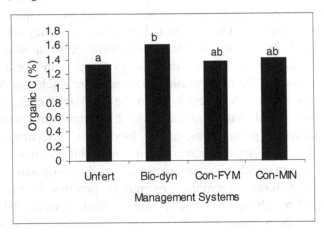

Figure 2 Microbial biomass carbon (a) and organic carbon (b) measurements from the DOK trial in Switzerland in 1997 (adapted from Fließbach *et al.* (2000)). Unfert = unfertilised; Bio-dyn = Biodynamic; Con-FYM = Conventional farm yard manure; Con-Min = Conventional with mineral fertiliser. Data in columns with similar numbers are not different.

Mäder *et al.* (2000) found that 30-60% more of the root length was colonised by arbuscular mycorrhizal (AM) fungi in plants grown in soils from the biodynamic and organic plots of the DOK trial compared with the conventionally managed plots. However, the highest values were measured in the unfertilised control. Approximately 50% of the variation of AM root colonisation was explained by chemical properties of the soils, particularly soluble P. The use of relatively insoluble reactive rock phosphate fertiliser on organic farms has also shown no decreased levels of AM infection in Australia (Ryan *et al.* 1994). Mäder *et al.* (2000) suggested factors such as fungicides and herbicides affected AM infection, directly by lowering spore viability and hyphal growth, and indirectly, by eliminating weeds which can act as hosts plants for AM fungi.

3.5   Roseworthy Farming Systems Trial - Australia

In 1989, a trial was established to assess the sustainability of 4 dryland broadacre farming systems under a Mediterranean climate (440 mm annual rainfall) on a sandy clay loam soil at the University of Adelaide, Australia (Penfold 1997). The conventional treatment used minimum tillage and direct drilling, gypsum, synthetic fertilisers and pesticides. This was compared with i) an organic rotation which used neomin and/or composted manure, green manure and mulches, and cultural and mechanical pest control; ii) biodynamic rotation which used neomin and/or composted manure, biodynamic preparations applied to the soil, cultural and mechanical pest control; and, iii) an integrated rotation which utilised minimum tillage and direct drilling, gypsum, municipal sludge and synthetic pesticides. After three and/or seven years there were no significant differences between systems in soil organic C (Penfold *et al.* 1995). There were also no differences with respect to microbial biomass, microbial activity, earthworm populations or infection of wheat roots by arbuscular mycorrhizae (AM) fungi. There was however significantly greater AM infection of medic (*Medicago truncatulata*) in the alternative systems compared with the conventional in 1996 (Table 5). This suggests that under dryland Mediterranean conditions, any changes in soil management will change biological fertility more slowly than in temperate climates.   The sustainability of European-designed organic or biodynamic farming systems under drier conditions may also be limited because of i) the inherently low nutrient and biological state of the soils, ii) distance from and inaccessibility to sources of organic matter inputs, iii) the destructive effects of tillage for weed control on fragile soils, and iv) climatic constraints. Although there are many examples of successful organic and biodynamic broadacre farmers in this environment, in general, sustainable systems under these conditions may require an integrated approach that incorporates the best of conventional and alternative farming systems (Rovira 1995). Extension of the principles underpinning a farming system to a different climatic zone without careful consideration may have disastrous consequences (Twomlow *et al.* 1999); local adaptation is critical.

Table 5. Arbuscular mycorrhizal (AM) infection of medic roots from Roseworthy Farming Systems field trial (adapted from Penfold 1997).

Management system	AM infection (%)*
Conventional	20a
Organic	37b
Biodynamic	34b
Integrated	35b

* Differences in letter denotes significant differences between means

3.6   Agroforestry - India

In 1983, a long term agroforestry experimental field site was established in a subtropical / semi arid climate at the CCS Haryana Agricultural University India. Shisham trees (*Dalbergia sissoo*) were planted in four spacing; no trees, 10 m x 10 m, 10 m x 5 m and 5 m x 5m. In spaces between rows wheat (*Triticum aestivum*) and cowpea (*Vigna sinensis*) have been grown in annual rotation since 1992 (Chander *et al.* 1998).

Chander *et al.* (1998) reported that by 1995 there was significantly greater organic C, total N, microbial biomass C, basal soil respiration and soil enzyme activity in treatments with tree-crop combinations than in the treatment without trees (Table 6). They also observed that soil organic matter, microbial biomass C and soil enzyme activities increased with decreasing tree spacing density (Table 6). However, Chander *et al.* (1998) also found that subsequent crop yields decreased by 16-17%, 23-27% and 42-66% with increasing tree spacing density compared with no trees. This emphasized that agronomically successful agroforestry systems must balance the competition between trees and crops for light and water, as well as increasing the nutrient inputs to the soil.

Agroforestry systems may enhance cycling of organic matter, decrease erosion losses, increase plant and faunal diversity and sequester more atmospheric C than comparable cropping systems (Sanchez *et al.* 1997). The positive impact of agroforestry systems on soil biological fertility has been attributed to the ameliorative effect of trees and the input of organic matter in leaf litter and fine roots, with significant differences between tree species due to the differences in amount and quality of litter (Kaur *et al.* 2000).

Table 6 Soil biological properties from soils under different agroforestry management systems for 12 years (adapted from Chander *et al.* (1998)).

Tree spacing	Organic C (%)	Total N (%)	Microbial biomass C (mg C kg⁻¹ soil)	[1]DH activity (µg TPF g⁻¹ soil 24 h⁻¹)[1]	[2]AP activity (µg PNP g⁻¹ soil h⁻¹)[2]
No trees	0.621a	0.069a	229a	27a	761a
10 x 10	0.624a	0.068a	235a	32b	792b
10 x 5	0.639b	0.072b	245b	36bc	854c
5 x 5	0.651c	0.073b	261c	39c	946d

[1] DH = Dehydrogenase activity   [2] AP = Alkaline phosphatase activity
Differences in letter denotes significant differences between means.

4.   CONCLUSION

This chapter has highlighted that the impact of agricultural systems on soil biological fertility depends on the extent to which they are 'microbial farming' i.e. managing the soil to the benefit of the microbial populations present in soil. Combinations of management practices which are locally adapted to maintain and increase soil organic matter status, and which optimise soil conditions to achieve maximal function of beneficial microbial populations, whilst minimising the impact of deleterious microorganisms, will deliver the most sustainable farming systems from the perspective of soil biological fertility. Consequently, any sustainable farming system, irrespective of its name, should include principles, recommendations or requirements which foster the development of such soil management strategies and practices and encourage their wide adoption.

5.   ACKNOWLEDGEMENTS

WRC would like to acknowledge the financial support of the New Zealand Foundation of Science and Technology. Rothamsted Research receives grant-in-aid form the UK Biotechnology and Biological Sciences Research Council.

6.   REFERENCES

Bloem J, Lebbink G, Zwart K B, Bouwman L A, Burgers S L G E, de Vos J A and de Ruiter P C 1994 Dynamics of microorganisms, microbiovores and nitrogen mineralisation under conventional and integrated management. Agriculture, Ecosystems and Environment 51: 129-143.

Bossio D A, Scow K M, Gunapala N and Graham K J 1998 Determinants of soil microbial communities: effects of agricultural management, management, season, and soil type on phospholipid fatty acid profiles. Microbial Ecology 36: 1-12.

Carpenter-Boggs L, Kennedy A C and Reganold J P 2000 (a) Organic and biodynamic management: effects on soil biology. Soil Science Society of America Journal 64: 1651-1659.

Carpenter-Boggs L, Reganold J P and Kennedy A C 2000 (b) Effects of biodynamic preparations on compost development. Biological Agriculture and Horticulture 17: 313-328.

Chander K, Goyal S, Nandal D P and Kappoor K K 1998 Soil organic matter, microbial biomass and enzyme activities in a tropical agroforestry system. Biology and Fertility of Soils 27: 168-172.

El Titi A, Boller E F and Gendrier J P 1993 Integrated production – Principles and technical guidelines. IOBC/WPRS Bulletin 16.

Fließbach A and Mäder P 1997 Carbon source utilisation by microbial communities in soilsunder organic and conventional farming practice. In: Microbial Communities – Functional versus structural approaches. Insam H and Rangger A (eds) pp 109-120. Springer, Berlin, Germany.

Fließbach A, Mäder P and Niggi U 2000 Mineralisation and microbial assimilation of ^{14}C-labelled straw in soils of organic and conventional agricultural systems. Soil Biology and Biochemistry 32: 1031-1039.

Gunapala N and Scow K M 1998 Dynamics of soil microbial biomass and activity in conventional and organic farming systems. Soil Biology and Biochemistry 30: 805-816.

Gunapala N, Venette R C, Ferris H and Scow K M 1998 Effects of soil management history on the rate of organic matter decomposition. Soil Biology and Biochemistry 30: 1917-1927.

Holland J M, Frampton G K, Çilgi T and Wratten S D 1994 Arable acronyms analysed- a review of integrated arable farming systems research in Western Europe. Annals of Applied Biology 125: 399-438.

International Federation of Organic Agriculture Movements, IFOAM 1998. Basic standards for organic production and processing. IFOAM. Tholey-Theley, Germany.

Kaur B, Gupta S R and Singh G 2000 Soil carbon, microbial activity and nitrogen availability in agroforestry systems on moderately alkaline soils in northern India. Applied Soil Ecology 15: 283-294.

Kennedy A C and Kremer R J 1996 Microorganisms in weed control strategies. Journal of Production Agriculture 9: 480-485.

Koepf H H, Pettersson B D and Schaumann W 1976 Biodynamic Agriculture. Anthroposophic Press. Spring Valley, NY. USA.

Letourneau D K and Goldstein B 2001 Pest damage and arthropod community structure in organic vs. conventional tomato production in California. Journal of Applied Ecology 38: 557-570.

Lundquist E J, Scow K M, Jackson L E, Uesugi S L and Johnson C R 1999 Rapid response of soil microbial communities from conventional, low-input, and organic farming systems to a wet/dry cycle. Soil Biology and Biochemistry 31: 1661-1675.

Mäder P, Edenhofer S, Boller T, Wiemken A and Niggli U 2000 Arbuscular mycorrhizae in a long-term field trial comparing low-input (organic, biological) and high-input (conventional) farming systems in a crop rotation. Biology and Fertility of Soils 31: 150-156.

Mäder P, Fließbach A, Dubois D, Gunst L, Fried P and Niggli U 2002 Soil fertility and biodiversity in organic farming. Science 296: 1694-1697.

Menalled F D, Gross K L and Hammond M 2001 Weed aboveground and seedbank community responses to agriculture management systems. Ecological Application 11: 1586-1601.

Nair P K R 1989 Agroforestry systems in the tropics. Kluwer, Boston.

National Research Council, NRC 1993 Sustainable agriculture and the environment in the humid tropics. National Academy Press. Washington DC. USA.

Oberson A, Besson J M, Maire N and Sticher H 1996 Microbiological transformations in soil - organic phosphorus transformations in conventional and biological cropping systems. Biology and Fertility of Soils 21: 138-148.

Penfold C M 1997 Biological Farming systems – Unravelling the mysteries. Final project report to GRDC. 16 pp.

Penfold C M, Miyan M S, Reeves T G and Grierson I T 1995 Biological farming for sustainable agricultural production. Australian Journal of Experimental Agriculture 35: 849-856.

Quimby P C, King L R and Grey W E 2002 Biological control as a means of enhancing the sustainability of crop/land management systems. Agriculture, Ecosystems and Environment 88: 147-152.

Reijntjes C, Haverkort B and Waters-Bayer A 1992 Farming for the future: an introduction to low-external input and sustainable agriculture. Macmillan Press. Basingstoke. UK.

Rovira A D 1995 Sustainable farming systems in the cereal-livestock areas of the Mediterranean region of Australia. *In*: Soil Management in Sustainable Agriculture. Cook H F and Lee H C (eds.) pp. 12-30. Proceedings of the Third International Conference on Sustainable Agriculture, 31 August – 4 September 1993, Wye College, University of London. Wye College Press. Ashford. UK.

Ryan M, Chilvers G and Dumaresq D 1994 Colonisation of wheat by VA-mycorrhizal fungi was found to be higher on a farm managed in an organic manner than on a conventional neighbour. Plant and Soil 60: 33-40.

Sanchez, P A, Buresh R J and Leakey R R B 1997 Trees, soils and food security Land Resources: On the edge of the Malthusian Precipice? Philosophical Transactions of the Royal Society of London B 352: 949-961.

Steiner R 1924 Agriculture: A course of eight lectures. Rudolph Steiner Press / Biodynamic Agriculture Association. London.

Stockdale E A, Lampkin N H, Hovi M, Keatinge R, Lennartsson E K M, Macdonald D W, Padel S, Tattersall F H, Wolfe M S and Watson C A 2001 Agronomic and environmental implications of organic farming systems. Advances in Agronomy 70: 261-327.

Temple S R, Friedman D B, Somasco O, Ferris H, Scow K and Klonsky K 1994 An interdisciplinary, experiment station-based participatory comparison of alternative crop management systems for California's Sacramento Valley. American Journal of Alternative Agriculture 9: 62-71.

Topp C F E, Stockdale E A, Fortune S, Watson C A and Ramsay S 2000 A comparison of a nitrogen budget for a conventional and an organic farming system. Aspects of Applied Biology. Farming Systems for the New Millennium. 62: 205-212.

Twomlow S, Riches C, O'Neill D, Brookes P and Ellis-Jones J 1999 Sustainable dryland smallholder farming in sub-saharan Africa. Annals of Arid Zone 38: 93-135.

Chapter 12

Sustainability of Soil Management Practices - a Global Perspective

Pete Smith[1] and David S. Powlson[2]

[1] *School of Biological Sciences, University of Aberdeen, Cruikshank Building, Aberdeen AB24 3UU, U.K.*
[2] *Agriculture and Environment Division, Rothamsted Research, Harpenden, Hertsfordshire, AL5 2JQ, U.K.*

1. INTRODUCTION

1.1 What is Sustainable Soil Management?

Sustainable development "meets the needs of the present without compromising the ability of future generations to meet their own needs for land" (W C E D 1987). We can adapt this definition to derive a definition for sustainable soil management: "soil management that meets the needs of the present without compromising the ability of future generations to meet their own needs from that soil". Thus, soil management is sustainable when it does not alter the capacity of the soil to provide for future needs. In this book, particular emphasis is placed upon the role of soil biology in the maintenance of soil sustainability. Management practices that threaten the soil biological community may also threaten soil sustainability by reducing the capacity of the soil to adapt in the future (Yachi and Loreau 1999).

L.K. Abbott & D.V. Murphy (eds) Soil Biological Fertility - A Key to Sustainable Land Use in Agriculture. 241-254.
© 2007 *Springer.*

Soil sustainability can be threatened by numerous management practices including over-cultivation, decreased or increased water abstraction, under-fertilisation or over-fertilisation, careless use of biocides, failure to maintain soil organic matter levels and clearing natural vegetation. These may threaten sustainability in a number of ways through physical and chemical processes (e.g. by increasing soil erosion, salinisation, desertification), or biological processes (e.g. by decreasing soil fertility). When soil management is poor, soil sustainability is often threatened by a combination of these factors at the same time.

Because the impacts of poor soil management are so severe in many areas of the world, the adoption of sustainable soil management is of crucial importance for the future of human and natural systems. Practices that improve the sustainable management of soils have already been described in detail in this book. In this chapter we consider soil sustainability from a global perspective and examine the current status of sustainable soil management and how it may change in the future.

2.   THE GLOBAL PERSPECTIVE – SOIL SUSTAINABILITY NOW AND IN THE FUTURE

2.1   Current Global Trends in Soil Sustainability

Trends in soil sustainability differ from region to region. In some of the more developed areas of the world, huge increases in yield per unit area over the last century (Amthor 1998, Matson *et al.* 1997), coupled with stabilising human population figures, have led to sustainable soil management becoming more prevalent than before. In Europe and much of the United States, unsustainable agricultural practices (such as those leading to the US mid-west 'dust bowl' of the 1930s; Cook *et al.* 1999) have given way to more sustainable forms of agriculture, including the conservation reserve program (Reeder *et al.* 1998, Lal *et al.* 1998) and other national equivalents. Improvements in crop varieties and a better knowledge of sustainable management practices mean that the pressure on land for agriculture in some of these areas is diminishing, either making land available for other uses such as woodland (Smith *et al.* 1997), or less intensive production systems (Lampkin 1990). In other areas of the world, such as the tropics, sustainable soil management may be more difficult, and slash-and-burn agriculture is still widely practised (Palm *et al.* 1996). For resource-poor farmers, short-term goals (such as the need to achieve a food crop this year in order to survive), will outweigh consideration of the longer-term impacts of soil management practices. Soil management practices are improving in these regions, but there is still a long way to go before their agricultural management systems can be regarded as sustainable (Palm *et al.* 1996, Sanchez 2000). Poverty in these regions leads to further land degradation (Barbier 2000) making soil sustainability even more difficult to achieve. Some authors have argued that the world's present development path is not sustainable, in that interactions between climate change, loss of biological diversity, increasing poverty and disease, and growing inequality, combine to increase the vulnerability of humans and nature (McCarthy and Dickson 2000). The development of sustainable soil management practices in the future has to be considered against the political, social and economic back-drop of the future world.

2.2   What is Likely to Happen to Soil Sustainability in the Future?

2.2.1   International agreements that may affect soil sustainability in the near future

A number of international agreements, conventions and instruments are likely to influence the management of soils in the near future. Article 3.4 of the Kyoto Protocol of the United Nations Framework Convention on Climate Change (UNFCCC), for example, explicitly mentions agricultural soils (http://www.unfccc.int/resource/docs/convkp/kpeng.pdf) for possible future inclusion as a biospheric sink for carbon (I P C C 2000a, b). The management practices that increase in soil organic carbon (e.g. reduced tillage, use of more organic amendments, greater use of mixed farming, inclusion of more wooded areas on farms, biofuel crop growth, protection from desertification etc.; Smith *et al.* 1997, 1998, Smith *et al.* 2000, Lal *et al.* 1998), may also help to improve soil biodiversity (e.g. Chan 2001), as well as soil sustainability, in what has been termed a "win-win strategy" (Lal *et al.* 1998). Other conventions, such as the United Nations Convention on Biological Diversity (http://www.biodiv.org/convention/), are also likely to yield benefits for soil sustainability through improved agricultural and land management (e.g. Liang *et al.* 2001). Other international and regional agreements, such as the United Nations Convention to Combat Desertification (UNCCD), United Nations Forum on Forests, and the Convention on Long-Range Trans-boundary Air Pollution (LRTAP), may also indirectly affect soil sustainability via constraints on land-management. In the following section, we give an example of how one such agreement, the Kyoto Protocol, might influence soil sustainability in the near future.

2.2.2   An example - possible impact of the Kyoto Protocol of soil sustainability via an impact on soil organic matter

Soil organic matter plays an important role in maintaining soil sustainability. It can help to maintain soil structure, retain soil moisture, prevent erosion, and can act as a reservoir for nutrients and as a source or sink for carbon (see Chapters 2, 3, and 4). Increasing levels of organic matter in soils is likely to enhance soil biological processes (see Chapters 2 and 3) and soil sustainability (Lal *et al.* 1998). In the Kyoto Protocol, agricultural soils are highlighted for possible future inclusion as a biospheric sink for carbon, which a party could use to help meet its CO_2 emission reduction targets. If agricultural soils were to be used as carbon sinks, there would be incentives to farmers and land managers to increase the soil organic carbon (SOC) content of their soils. In this example, we examine the global potential to increase SOC, and thereby gain some insight into the extent to which soil sustainability might be improved in the future.

There is considerable potential to increase soil carbon stocks due to the abundance of agricultural soils which are depleted in carbon. Cultivation has resulted in a loss of 55 Pg (10^{15}g) carbon from soils world-wide (Cole *et al.* 1996). The best agricultural management practices have the potential to restore some of this soil carbon. At the global scale, Cole *et al.* (1996), estimated that 0.4-0.8 Pg carbon could be sequestered in agricultural soils each year, corresponding to an increase of 40-80 Pg carbon over 100 years. These figures are however considered a little high (see below).

In certain areas, some historical carbon loss from agricultural soils may have already been reversed. For example, the introduction of conservation tillage in the USA is estimated to have increased SOC stocks by about 1.4 Pg over the past 30 years (Donigian *et al.* 1994), with the potential to store a further 5 Pg over the next 50 years (Kern and Johnson 1993, Lal *et al.* 1998).

Over the next 50-100 years, if one includes all available management practices, some estimates indicate that there is the potential to sequester 0.075 to 0.208 Pg carbon per year in US arable land (Lal *et al.* 1998, Metting *et al.* 1999), which is equivalent to 7.5-20.8 Pg carbon over 100 years. In the USA, it has been estimated that full adoption of best management practices could restore SOC levels to about 75-90% of their pre-cultivation levels (Donigian *et al.* 1994). Figures from Europe are of a similar order, i.e. 0.113 Pg carbon could be offset per year (including carbon offsets from bioenergy crops planted on surplus arable land; Smith *et al.* 2000). Over 100 years this is equivalent to a carbon offset of 11.3 Pg. In some regions, then, there is the potential to restore much of the historic carbon lost though cultivation, but globally the potential is lower. Much of the potential to reverse SOC loss occurs in temperate regions such as Europe and North America. The figures for USA of Lal *et al.* (1998) and for Europe of Smith *et al.* (2000) suggest that, at best, about 1/3 to 1/2 of the SOC lost through agriculture globally, can be restored over the next 100 years.

Yearly increases in SOC can be sustained for only 50-100 years (Smith *et al.* 1997, Cole *et al.* 1996); as ecosystems reach a new equilibrium position, yearly SOC increases slow and eventually cease. In this context, for activities adopted under Kyoto article 3.4, soil carbon sink saturation will occur within about 50-100 years of adoption. Although not calculated on an area basis, if 1/3 to 1/2 of SOC lost through agriculture can be restored, soil sustainability might increase on a similar proportion of current agricultural land.

Land uses that could enhance carbon sequestration and improve soil sustainability in terrestrial ecosystems include: i) some agricultural land uses (see above), ii) biomass crops, grassland, rangeland and forest, iii) the protection and creation of wetlands, urban forest and grassland, iv) the manipulation of deserts and degraded lands, and v) the protection of sediments, aquatic systems, tundra and taiga (Metting *et al.* 1999). Over a 50 year time period, the total terrestrial carbon sequestration potential, including all of these activities is estimated to be 5.65-8.71 Pg carbon per year (Metting *et al.* 1999), roughly 10 times the carbon sequestration potential of agricultural land alone. Given the importance of soil organic matter in maintaining soil biological fertility and sustainability, this level of carbon sequestration would almost certainly lead to improved soil sustainability in many regions of the world over the next 50 years. However, the word 'potential' needs to be emphasised when considering carbon sequestration in soil. In order to achieve these increases in soil carbon content, it would be necessary to achieve the maximum uptake of various land management practices which in reality, may be difficult. It was largely for this reason that a cautious approach was taken to the role land carbon sinks may have in mitigating climate change in a recent study by the Royal Society (Royal Society 2001).

2.2.3 Soil sustainability over the next 50-100 years

Although the conventions and instruments mentioned in 2.2.1 and 2.2.2 (this Chapter) will influence soil sustainability in the near future, long-term trends will be constrained by social, political and economic growth in different regions of the world. While the political and economic future cannot be forecast, a number of possible futures can be examined. In its Special Report on (greenhouse gas) Emission Scenarios (SRES), the IPCC (2000b) developed a range of "story-lines" on future global development. The SRES story lines have been adopted for use in climate change impact studies (e.g. Parry 2000). The interpretation of the SRES story lines as used in Parry (2000) are summarised schematically in Figure 1. The x-axis represents the spatial scale of the political units, ranging from local, devolved government to a high level of global co-operation. The y-axis represents the economic framework, ranging from a highly individualistic free-market economy to a collective, more 'communitarian' economy. The four quarters of the figure represent different global scenarios. For the purposes of emission scenarios, A1 was further divided into three scenarios depending upon fossil fuel use, but that subdivision is less relevant here.

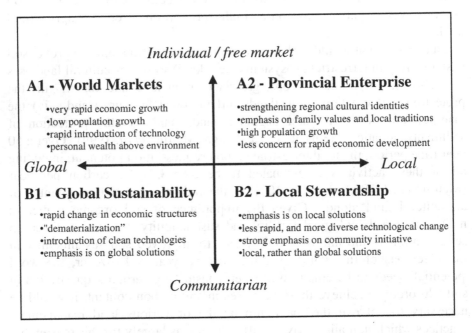

Figure 1 Schematic representation of the story lines used in developing the IPCC greenhouse gas emission scenarios.

2.2.4   What are the impacts of these possible futures upon soil sustainability?

World markets (A1). Under this scenario the emphasis would be on the pursuit of economic growth rather than sustainable development, and policies would be based on the assumption that natural systems are largely resilient to human stress (Parry 2000). Under the world markets scenario, soil sustainability could suffer due to a lack of concern for developing sustainable land management practices. Furthermore, although rising income levels will make everyone richer, the richest will become relatively richer compared to the poorest (Parry 2000). Poverty is well known to drive deforestation (Weis 2000) and to lead to land degradation (Barbier 2000). Increasing the income of the poorest people in the world would only result in improved soil sustainability if poverty were alleviated by the trickle-down economics engendered in this scenario.

Global Sustainability (B1).   Under this scenario there would be an increasing tendency to find global solutions to global problems through strong global institutions and the adoption of international regulation (Parry 2000). International trade would be strong, but environmental requirements would have primacy in the case of a conflict. There is also predicted to be significant government commitment to making agriculture more environmentally friendly (OECD 1996).   Soil sustainability would undoubtedly be improved under this scenario with global regulation and intervention to improve soil sustainability in the developing world, as well as in the developed world.

Provincial Enterprise (A2). Under this scenario policy decisions would be taken at the national and sub-national level, and society would be organised according to short-term consumerist values.   Protectionist economic and trade policies would constrain innovation and stifle economic development in poorer countries. A greater economic gulf would develop between the developed northern countries and poorer southern countries.   Politicians would regard environmental quality as a low priority compared to protecting the national economy and consumer demand for growth (Parry 2000). Soil sustainability under this scenario would suffer in developed countries, but would suffer even more in less-developed regions, where soils are already more vulnerable.

Local Stewardship (B2).   Under this scenario environmental problems would be solved locally and according to local conditions.   Governance would occur at the local level and would be more eco-centric in nature. The world would become more heterogeneous and relative inequality may arise due to lack of co-ordinated regional action. The environment would benefit under this scenario, but not as much as under the Global Sustainability *(B1),* since environmental policy would lack spatial co-ordination (Parry 2000). Soil sustainability under this scenario would also improve, provided that resources were available in the poorest areas to implement local sustainable land-management policies.

Among these scenarios, human population growth would be lowest under the Global Markets and Global Sustainability scenarios.   These scenarios would therefore create less pressure for new agricultural land than would the Local Stewardship and Provincial Enterprise scenarios, and would thus be less likely to threaten global soil sustainability. Taking human population growth into account, the least favourable future for soil sustainability would be Provincial Enterprise and the most favourable would be Global Sustainability. Whilst Local Stewardship might produce a more environmentally-conscious world, the higher population compared to the Global Markets scenario would trade off eco-centric land-management policies against the need for more land for food production, in turn leading to an increased threat of soil degradation.

Climate change may further increase the threat to soil sustainability in poorer countries because cereal crop yields are predicted to decline in most tropical and sub-tropical regions under future climates (Rosenzweig and Parry 1994, Fischer *et al.* 2001), in countries which have a low capacity to adapt (IPCC 2001). The decrease in organic matter returns to the soil, and the increased need for new land to counteract the effects of lower yields, would enhance negative impacts on soil sustainability in these regions. Even in some temperate regions there is increased risk of crop failure caused by drought, resulting from greater variation in weather from year to year even if the average change in climate is small (Porter and Semenov 1999). The likely overall impacts on soil sustainability of each of the future world story-lines is summarised in Figure 2.

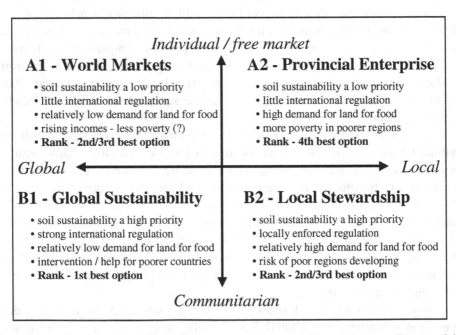

Individual / free market

A1 - World Markets
- soil sustainability a low priority
- little international regulation
- relatively low demand for land for food
- rising incomes - less poverty (?)
- **Rank - 2nd/3rd best option**

A2 - Provincial Enterprise
- soil sustainability a low priority
- little international regulation
- high demand for land for food
- more poverty in poorer regions
- **Rank - 4th best option**

Global ←——————————————————→ *Local*

B1 - Global Sustainability
- soil sustainability a high priority
- strong international regulation
- relatively low demand for land for food
- intervention / help for poorer countries
- **Rank - 1st best option**

B2 - Local Stewardship
- soil sustainability a high priority
- locally enforced regulation
- relatively high demand for land for food
- risk of poor regions developing
- **Rank - 2nd/3rd best option**

Communitarian

Figure 2 Summary of likely impacts on soil sustainability of the story-lines used in developing the IPCC emission scenarios.

As it is not possible to predict exactly what will happen to soil sustainability in the future, here we use accepted, realistic story-lines, which are based on credible socio-economic reasoning and modelling, to assess the impacts of a range of possible futures. Some possible futures are perhaps more likely than others, and recent political / economic trends have been towards globalisation and the pursuit of increased free trade (McCarthy and Dickson 2000). However, a number of international conventions and instruments are currently in place that are likely to improve soil sustainability in the near future. The future, at least regionally, is likely to contain components of all scenarios, but based upon recent trends, the future will probably most closely resemble a combination of the Global Market and Global Sustainability scenarios. In order to cope with the uncertainty inherent when planning for the future, soil management strategies should be developed which are beneficial now, and which are predicted to also be beneficial in the future.

3.   DEVELOPING A 'NO REGRETS' STRATEGY FOR SUSTAINABLE SOIL MANAGEMENT

A move towards globally sustainable soil management needs to be coupled with solutions to the other related environmental problems and socio-economic / political issues such as poverty, food security and over-population. Global solutions to soil sustainability will be best achieved as part of larger, comprehensive, sustainable development strategies. However, there are a number of best-management practices that could be adopted now to improve soil sustainability in the future, irrespective of the political and economic landscape of the future. These practices include the cessation of clearing natural vegetation, maintenance of soil organic matter via careful tillage, plant residue management, use of organic amendments, fertilisation and careful irrigation, the careful choice of crops and rotations and the careful use of biocides and soil amendments. These practices are similar to those described in Chapter 11 in which a move towards 'microbial farming' is advocated.

The strategy, whereby management practices known to benefit the soil now are implemented, with the expectation that it will improve the resilience of these soils to future challenges, is known as a 'no regrets' strategy. A schematic representation of how a 'no regrets' policy might increase the resilience of the soil is shown in Figure 3.

The main plot of each part of Figure 3 (adapted from a plot used to demonstrate ecosystem vulnerability in Smit *et al.* 2000) shows the variation in the value of an environmental stress. While the stress falls within the coping zone (shaded), the soil / ecosystem is resilient to the environmental stress. Problems occur when the value falls outside this zone. In Figure 3-a, the system management remains the same giving the same coping range throughout the period, whereas in Figure 3-b, the soil management is changed at time-zero to improve the resilience of the system.

a)

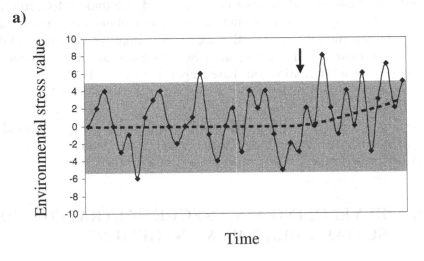

b)

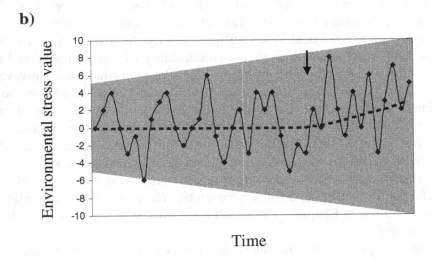

Figure 3 Environmental stress, coping ranges and improved soil management adapted from Smit *et al.* (2000). The main plot of each part shows the variation in an environmental stress. The broken line shows the mean value of that soil parameter over time. The arrow shows the onset of an environmental change. The shaded area shows the coping range (Smit *et al.* 2000). In a), the system management remains the same, giving the same coping range throughout the period. In b), the soil management is changed at time zero to improve the resilience of the system. See text for further explanation.

As seen in Figure 3-a, the stress on the soil increases after the environmental change (marked by the arrow) and moves outside the coping range three times in the last quarter of the time period (having strayed outside of the coping range twice in the first 3/4 of the period). However, when sustainable soil management is implemented at time zero (Figure 3-b), the coping range of the system increases such that the environmental stress falls within the expanding coping range at all times after the environmental change, and is only once outside the range before the environmental change. Figure 3 demonstrates schematically that a 'no regrets' policy implemented now, will increase the resilience of the soil to environmental stresses in the present (before an environmental change) as well as in the future.

4. CONCLUDING REMARKS

In the near future, a number of current international agreements and instruments such as the Kyoto Protocol, may indirectly improve soil sustainability by providing incentives to better manage soil organic matter. It is possible that 1/3 to 1/2 of the soil carbon lost globally through agriculture might be restored over the next 100 years if the Kyoto Protocol is fully implemented. If such an investment in improved soil management were made, soil sustainability would almost certainly improve in many regions of the world in the near future.

Soil sustainability in the more distant future will depend upon the world's future social, economic and political development. One scenario (Global Sustainability) would clearly benefit soil sustainability, one (Provincial Enterprise) would clearly damage it, and the other two (Global Markets and Local stewardship) would provide mixed incentives / disincentives for sustainable soil management.

At the regional scale, the education of farmers, land-managers and regional planners would also help to enhance soil sustainability, since the implementation of known technology is currently hampered by education (Sanchez 2000).

At the global scale, soil sustainability needs to be tackled in hand with other related problems. The IPCC (2001) noted that global, regional and local environmental issues such as climate change, loss of biodiversity, desertification, stratospheric ozone depletion, regional acid deposition and local air quality are inextricably linked. The lack of soil sustainability should certainly be added to that list. The IPCC (2001) further noted that recognising the linkages among local, regional and global environmental issues, and their relationship to meeting human needs, provides an opportunity to address global environmental issues at the local, national and regional level in an integrated manner that is cost-effective and meets sustainable development objectives. The importance of integrated approaches to sustainable environmental management is becoming ever clearer.

In addition to attempting to solve a raft of environmental problems together, social and economic problems also need to be addressed in the same package. All of the scientific and technical measures outlined in this book have the potential to improve soil sustainability, but the extent to which soil management becomes sustainable will be determined by how widely these measures are implemented. Soil sustainability is ultimately a human problem and it is impossible to separate soil sustainability from the political landscape of the future. Threats to soil sustainability (and a range of other problems) would be reduced by control of the human population size as this would ease the pressure on land for food production. Poverty remains the main driver for soil degradation in the poorest parts of the world, where the soil is most vulnerable (Barbier 2000). At the global scale, relief of poverty in these regions would probably do more to improve soil sustainability than any of the scientific or technical measures described. The development of sustainable soil management practices clearly requires a multidisciplinary approach and an awareness by all involved of the multiple goals to be achieved. It can be seen as a part of a wider international initiative termed "sustainability science". This is an attempt to bring together the often conflicting efforts to meet fundamental human needs while preserving the life-support systems of planet Earth (http://sustainabilityscience.org; http://sustsci.harvard.edu/keydocs.htm#sustsci; Kates *et al.* 2001;).

It is clear that the political and economic landscape of the future will determine the feasibility of many strategies to promote sustainable soil management, but there are a number of best management practices available now that could be implemented to improve soil sustainability now, and in the future (a no regrets policy). Since these practices are consistent with, and may even be encouraged by, many current international agreements and conventions, their rapid adoption should be encouraged as widely as possible.

5.   ACKNOWLEDGEMENTS

Rothamsted Research receives grant-in-aid form the UK Biotechnology and Biological Sciences Research Council.

6. REFERENCES

Amthor J S 1998 Perspective on the relative insignificance of increasing atmospheric CO_2 concentration to crop yield. Field Crops Research 58: 109-127.

Barbier E B 2000 The economic linkages between rural poverty and land degradation: some evidence from Africa. Agriculture, Ecosystems and Environment 82: 355-370.

Chan K Y 2001 An overview of some tillage impacts on earthworm population abundance and diversity - implications for functioning in soils. Soil and Tillage Research 57: 179-191.

Cole V *et al.*. 1996 Agricultural options for mitigation of greenhouse gas emissions. *In:* Impacts, Adaptations and Mitigation of Climate Change: Climate Change 1995. R T Watson, M C Zinyowera, R H Moss and D J Dokken (eds.) pp. 745-771. Scientific-Technical Analyses. Cambridge University Press. New York.

Cook E R, Meko D M, Stahle D W and Cleaveland M K 1999 Drought reconstructions for the continental United States. Journal of Climate 12: 1145-1162.

Donigian A S (Jr), Barnwell T O (Jr.), Jackson R B (IV), Patwardhan A S, Weinrich K B, Rowell A L, Chinnaswamy R V and Cole C V 1994 Assessment of Alternative Management Practices and Policies Affecting Soil Carbon in Agroecosystems of the Central United States. p. 194. US EPA Report EPA/600/R-94/067. Athens.

Fischer G, Shah M, van Velthuizen H and Nachtergaele F O 2001 Executive Summary Report: Global Agro-ecological Assessment for Agriculture in the 21st Century. International Institute for Applied Systems Analysis. Laxenburg, Austria.

I P C C 2000 (a). Special Report on Land Use; Land-Use Change and Forestry. p. 377. Cambridge University Press. Cambridge, UK.

I P C C 2000 (b). Special Report on Emission Scenarios. p. 570. Cambridge University Press. UK.

I P C C 2001 Climate Change 2001: The Scientific Basis. p. 944. Cambridge University Press. UK.

Kates R W, Clark W C, Corell R, Hall J M, Jaeger C C, Lowe I, McCarthy J J, Schellnhuber H J, Bolin B, Dickson N M, Faucheux S, Gallopin G C, Grübler A, Huntley B, Jäger J, Jodha N S, Kasperson R E, Mabogunje A, Matson P, Mooney H, Moore B, O'Riordan T and Svedin U 2001 Sustainability Science. Science 292: 641-642.

Kern J S and Johnson M G 1993 Conservation tillage impacts on national soil and atmospheric carbon levels. Soil Science Society of America Journal 57: 200-210.

Lal R, Kimble J M, Follet R F and Cole C V 1998 The potential of U.S. cropland to sequester carbon and mitigate the greenhouse effect. p. 128. Ann Arbor Press. Chelsea, M I.

Lampkin N 1990 Organic Farming. Farming Press. Ipswich, UK.

Liang L, Stocking M, Brookfield H and Jansky L 2001 Biodiversity conservation through agrodiversity. Global Environmental Change 11: 97-101.

Matson P A, Parton W J, Power A G and Swift M J 1997 Agricultural intensification and ecosystem properties. Science 277: 504-509.

McCarthy J J and Dickson N M 2000 From Friibergh to Amsterdam: On the road to sustainable science. I G B P Newsletter 44: 6-8.

Metting F B, Smith J L and Amthor J S 1999 Science needs and new technology for soil carbon sequestration. *In:* Carbon Sequestration in Soils: Science, Monitoring and Beyond. N J Rosenberg, R C Izaurralde and E L Malone (eds.) pp. 1-34. Battelle Press. Columbus, Ohio.

O E C D 1996 Environmental performance in the 1990s: Progress in the 1990s. O E C D. Paris.

Palm C A, Swift M J and Woomer P L 1996 Soil biological dynamics in slash-and-burn agriculture. Agriculture, Ecosystems and Environment 58: 61-74.

Parry M (ed.) 2000 Assessment of potential effects and adaptations for climate change in Europe: The Europe ACACIA Project. p. 320. Jackson Environment Institute, University of East Anglia. Norwich, UK.

Porter J R and Semenov M A 1999 Climatic variability and crop yields in Europe (Scientific Correspondence). Nature 400: 724.

Reeder J D, Schuman G E and Bowman R A 1998 Soil C and N changes on conservation reserve program lands in the Central Great Plains. Soil and Tillage Research 47: 339-349.

Rosenzweig C and Parry M 1994 Potential impact of climate change on world food supply. Nature 367: 133-138.

Sanchez P A 2000 Linking climate change research with food security and poverty reduction in the tropics. Agriculture, Ecosystems and Environment 82: 371-383.

Smit B, Burton I, Klein R J T and Wandel J 2000 An anatomy of adaptation to climate change and variability. Climatic Change 45: 223-251.

Smith P, Powlson D S, Glendining M J and Smith J U 1997 Potential for carbon sequestration in European soils: preliminary estimates for five scenarios using results from long-term experiments. Global Change Biology 3: 67-79.

Smith P, Powlson D S, Glendining M J and Smith J U 1998 Preliminary estimates of the potential for carbon mitigation in European soils through no-till farming. Global Change Biology 4: 679-685.

Smith P, Powlson D S, Smith J U, Falloon P and Coleman K 2000 Meeting Europe's climate change commitments: Quantitative estimates of the potential for carbon mitigation by agriculture. Global Change Biology 6: 525-539.

The Royal Society 2001 The role of land carbon sinks in mitigating global climate change. pp. 27. Policy Document 10/01. The Royal Society, www.royalsoc.ac.uk London, UK.

W C E D (World Commission on Environment and Development) 1987 Our Common Future. p. 383. Oxford University Press. Oxford, UK.

Weis T 2000 Beyond peasant deforestation: environment and development in rural Jamaica. Global Environmental Change 10: 299-305.

Yachi S and Loreau M 1999 Biodiversity and ecosystem productivity in a fluctuating environment; The insurance hypothesis. Proceedings of the National Academy of Sciences of the United States of America 96: 1463-1468.

LIST OF CONTRIBUTORS

Abbott, Lynette

School of Earth and Geographical Sciences, Faculty of Natural and Agricultural Sciences, The University of Western Australia, Crawley, WA, 6009, Australia.
Email: labbott@cyllene.uwa.edu.au

Brookes, Philip

Agriculture and Environment Division, Rothamsted Research, Harpenden, Hertfordshire, AL5 2JQ, United Kingdom.
Email: philip.brookes@bbsrc.ac.uk

Chan, Yin

Wagga Wagga Agricultural Institute, NSW Agriculture, PMB, Wagga Wagga, NSW 2650, Australia.
Email: yin.chan@agric.nsw.gov.au

Clapperton, Jill

Agriculture and Agri-Food Canada, Lethbridge Research Centre, P. O. Box 3000, Lethbridge Alberta T1J 4B1, Canada.
Email: clapperton@agr.gc.ca

Cookson, William

School of Earth and Geographical Sciences, Faculty of Natural and Agricultural Sciences, The University of Western Australia, Crawley, 6009, Australia.
Email: wcookson@cyllene.uwa.edu.au

Douds, David Jr.

USDA-ARS, Eastern Regional Research Centre, 600 E. Mermaid Lane, Wyndmoor, PA 19038, United States of America.
Email: ddouds@ars.usda.gov

Goulding, Keith

Agriculture and Environment Division, Rothamsted Research, Harpenden, Hertfordshire, AL5 2JQ, United Kingdom.
Email: keith.goulding@bbsrc.ac.uk

Gupta, V.V.S.R.

CSIRO Land and Water, Waite Campus, PMB No. 2, Glen Osmond, 5064, SA, Australia.
Email: Gupta.Vadakattu@csiro.au

255

Haq, Krystina

School of Earth and Geographical Sciences, Faculty of Natural and Agricultural Sciences, The University of Western Australia, Crawley, WA, 6009, Australia.
Email: Krys.Haq@uwa.edu.au

Hendrix, Paul

University of Georgia, Institute of Ecology and Department of Crop and Soil Sciences, Athens, GA, United States of America.
Email: hendrixp@uqa.edu

Johnson, Nancy Collins

Northern Arizona University, Environment Sciences and Biological Sciences, P.O. Box 5694, Flagstaff, AZ 86001, United States of America.
Email: Nancy.Johnson@nau.edu

Larney, Frank

Agriculture and Agri-Food Canada, Lethbridge Research Centre, P. O. Box 3000, Lethbridge Alberta T1J 4B1, Canada.
Email: larney@agr.gc.ca

Marschner, Petra

Soil and Land Systems, School of Earth and Environmental Sciences, The University of Adelaide, Waite Campus, PMB1, Glen Osmond, SA, 5064, Australia.
Email: petra.marschner@adelaide.edu.au

McInnes, Alison

School of Environment and Agriculture, College of Science, Technology and Environment, University of Western Sydney-Hawkesbury, Locked Bag 1797, Penrith South DC, NSW, 1797, Australia.
Email: a.mcinnes@uws.edu.au

Murphy, Daniel

School of Earth and Geographical Sciences, Faculty of Natural and Agricultural Sciences, The University of Western Australia, Crawley, WA, 6009, Australia.
Email: dmurphy@cyllene.uwa.edu.au

Osler, Graham

The Macaulay Land Use Research Institute, Craigiebuckler, Aberdeen, AB15 8QH, United Kingdom.
Email: G.Osler@macaulay.ac.uk

Powlson, David

Agriculture and Environment Division,
Rothamsted Research, Harpenden, Hertfordshire,
AL5 2JQ, United Kingdom.
Email: david.powlson@bbsrc.ac.uk

Rengel, Zdenko

School of Earth and Geographical Sciences,
Faculty of Natural and Agricultural Sciences, The
University of Western Australia, Crawley, WA,
6009, Australia.
Email: zrengel@fnas.uwa.edu.au

Sivasithamparam, K.

School of Earth and Geographical Sciences,
Faculty of Natural and Agricultural Sciences, The
University of Western Australia, Crawley, WA,
6009, Australia.
Email: siva@cyllene.uwa.edu.au

Smith, Pete

School of Biological Sciences,
University of Aberdeen, Cruikshank Building,
Aberdeen, AB24 3UU, United Kingdom.
Email: pete.smith@abdn.ac.uk

Stockdale, Elizabeth

Agriculture and Environment Division,
Rothamsted Research, Harpenden, Hertfordshire,
AL5 2JQ, United Kingdom.
Email: E.A.Stockdale@newcastle.ac.uk

van Vliet, Petra

Wageningen University,
Sub-department of Soil Quality, Wageningen, The
Netherlands.
Email: Petra.vanVliet@wur.nl

Wardle, David

Landcare Research,
P.O. Box 69,
Lincoln, New Zealand
Email: wardled@landcareresearch.co.nz

Williamson, Wendy

Landcare Research, P. O. Box 69,
Lincoln, New Zealand.
Email: WilliamsonW@landcareresearch.co.nz

INDEX